Advances in Boiling and Condensation

Edited by Igor L. Pioro

Published in London, United Kingdom

Advances in Boiling and Condensation
http://dx.doi.org/10.5772/intechopen.104037
Edited by Igor L. Pioro

Contributors
Aniket M. Rishi, Armando Nava-Dominguez, Yasuo Koizumi, Sanja Milivojevic, Vladimir Stevanovic, Milan M. Petrovic, Milica Ilic, Igor Leonardovich Pioro

First published in London, United Kingdom, 2024 by IntechOpen
IntechOpen is the global imprint of INTECHOPEN LIMITED, registered in England and Wales, registration number: 11086078, 5 Princes Gate Court, London, SW7 2QJ, United Kingdom

British Library Cataloguing-in-Publication Data
A catalogue record for this book is available from the British Library

Additional hard and PDF copies can be obtained from orders@intechopen.com

Advances in Boiling and Condensation
Edited by Igor L. Pioro
p. cm.
Print ISBN 978-1-80356-293-3
Online ISBN 978-1-80356-294-0
eBook (PDF) ISBN 978-1-80356-295-7

For EU product safety concerns:
IN TECH d.o.o., Prolaz Marije Krucifikse Kozulić 3, 51000 Rijeka, Croatia,
info@intechopen.com or visit our website at intechopen.com.

Meet the editor

Professor Igor Pioro, Ph.D., DSc (Tech), is a professional engineer, foreign fellow of the National Academy of Sciences, Ukraine, and fellow of the American Society of Mechanical Engineers (ASME), Canadian Academy of Engineering, Canadian Society of Mechanical Engineers, and Engineering Institute of Canada. He is a lifetime member of the American Nuclear Society, and a member of Canadian NS. Dr. Pioro is an internationally recognized scientist in nuclear engineering (thermal hydraulics of nuclear reactors, Generation-IV reactors, etc.) and thermal sciences/engineering (boiling, forced convection including supercritical pressures, etc.). He is the author/co-author of more than 535 publications, including books, book chapters, journal articles, conference proceedings, and major technical reports. He also holds twenty-six patents. Dr. Pioro is the founding editor and editor-in-chief of the ASME *Journal of Nuclear Engineering and Radiation Science*. In 2011–2012, he was the chair of the Executive Committee of the ASME Nuclear Engineering Division and the chair of the International Conference On Nuclear Engineering (ICONE-20).

Contents

Preface

Advances in Boiling and Condensation is written by well-known experts from top universities and research organizations in Canada, Japan, Serbia, and the United States. The book consists of five chapters, three of which are dedicated to the phenomena of boiling heat transfer, one that covers both boiling and condensation and one on condensation heat transfer.

Boiling and condensation are two types of convection heat transfer with phase change, characterized by high heat-transfer coefficients within the range of 2500–100,000 W/m^{2}K. Boiling and condensation are widely used in various industries. However, possibly the largest and most significant application is in the power industry, where they are used in thermal as well as nuclear power plants, which are equipped with subcritical-pressure Rankine steam-turbine power cycles. Even supercritical-pressure coal-fired power plants still rely on condensation of steam from low-pressure turbines.

In general, boiling is a heat-transfer process during which vapor bubbles are created on a heated surface (nucleate boiling) or inside an overheated liquid (bulk boiling). Boiling has been used by humans for tens of thousands of years for cooking, however, its application in industry started sometime in the 17th century. Moreover, actual research into boiling heat-transfer phenomena began only around the 1920s. Several major types of boiling processes can be identified. These include natural-convection pool boiling vs. forced-convection flow boiling and nucleate boiling vs. bulk boiling. Major nucleate pool-boiling characteristics are Onset of Nucleate Boiling (ONB), Heat Transfer Coefficient (HTC), Critical Heat Flux (CHF), HTC at film pool boiling, minimum heat flux at film pool boiling, and HTC at transition boiling. Quite similar characteristics correspond to flow boiling, which are the Onset of subcooled Nucleate Boiling (ONB), Onset of Significant Void (OSV), HTC, CHF, and Post-Dryout (PDO) heat transfer. Despite more than 100 years of active research and many years of applications, boiling phenomena/heat transfer are still not fully investigated and understood. There have been some attempts to develop boiling-phenomena theories, but, unfortunately, they are not yet practical. Therefore, more or less all practical calculations of various boiling characteristics/parameters rely heavily on empirical correlations that were obtained experimentally.

Chapter 1 of this book provides a summary of the latest developments in nucleate pool boiling and flow boiling, the latter related to boiling inside circular flow geometries or inside bundle/fuel-rod assemblies on fuel rods. Chapter 2 focuses on heat transfer and hydraulic resistance in fuel bundles of nuclear-power reactors, mainly, Pressurized Heavy Water Reactors (PHWRs), particularly, CANDU reactors. However, boiling heat-transfer experiments are usually performed with light water instead of heavy water (both fluids have quite similar thermophysical/thermodynamics properties), therefore, this chapter is also useful for our understanding of heat-transfer specifics in Pressurized light-Water Reactors (PWRs).

Chapter 3 examines boiling heat-transfer enhancement with graphene-based functional coatings. Pool-boiling heat transfer has proven to be the most effective way to dissipate high heat fluxes and achieve efficient cooling in many industrial applications, including high-power-electronics cooling, data-center cooling, heat exchangers, batteries, refrigeration, and air conditioning. Graphene, with its high thermal conductivity, has been implemented in numerous studies for improving both the CHF and HTC in pool-boiling heat transfer. This chapter introduces various graphene-based nanomaterials and basics related to the structure and characterization of graphene. It also highlights notable research work on graphene-based coatings for pool-boiling enhancements.

Chapter 4 is dedicated to water hammer in two-phase systems, which is induced by direct steam condensation on subcooled water or by separation of the subcooled water column. This results in the most intensive pipeline pressure surges. Amplitudes of pressure spikes along the course of these dangerous transients strongly depend on the condensation and evaporation rates. This chapter provides a literature review of thermal-hydraulic models for the prediction of water-hammer phenomenon in two-phase systems. Available water-hammer experimental conditions were numerically simulated with the new modelling approach.

Finally, Chapter 5 highlights a special case of heat transfer during condensation. In general, there are two main modes of condensation: film and dropwise condensation. However, some other special cases exist, one of which is steam condensation on a water jet in a steam injector. This chapter studies the heat-transfer mechanism within such conditions.

We hope that you will find this book a useful and informative resource. Please enjoy reading our new book!

On behalf of all authors,
Editor of the book,

Igor L. Pioro
Professor,
Foreign Fellow of National Academy of Sciences of Ukraine,
Ukraine

Founding Editor and Editor-in-Chief,
ASME Journal of Nuclear Engineering and Radiation Science

Department of Energy and Nuclear Engineering,
Faculty of Engineering and Applied Science,
Ontario Tech University (University of Ontario Institute of Technology),
Oshawa, Canada

Chapter 1

Advances and Challenges of Boiling Heat Transfer

Igor L. Pioro

Abstract

Boiling is a heat-transfer process during which vapor bubbles are created on a heated surface (nucleate boiling) or inside overheated liquid (bulk boiling). Boiling has been used by humans for tens of thousands of years for cooking, however, its application in industry started somewhere in the seventeenth century. Moreover, actual research into boiling-heat-transfer phenomena started only around 1920s. In general, several major types of boiling process can be identified: natural-convection pool boiling vs. forced-convection flow boiling and nucleate boiling vs. bulk boiling. Major nucleate-pool-boiling characteristics are as the following: Onset of Nucleate Boiling (ONB); Heat Transfer Coefficient (HTC); Critical Heat Flux (CHF); HTC at film pool boiling; minimum heat flux at film pool boiling; and HTC at transition boiling. Quite similar characteristics correspond to flow-boiling: Onset of subcooled Nucleate Boiling (ONB); Onset of Significant Void (OSV); HTC; CHF; and Post-DryOut (PDO) heat transfer. In spite of more than 100 years of active research and many years of applications, boiling phenomena/heat transfer are still not fully investigated and understood. There are some attempts to develop boiling-phenomena theories, but, unfortunately, they are not so practical yet. Therefore, more or less all practical calculations of various boiling characteristics/parameters rely heavily on empirical correlations, which were obtained experimentally. Due to this sophisticated studies are performed into boiling phenomena in the world.

Keywords: pool boiling, flow boiling, nucleate boiling, heat transfer coefficient, critical heat flux

1. Introduction: History notes

Based on various sources (Wikipedia, 2023), there is some evidence that ancient humans have started to boil water as early as ~30,000 years ago during the Upper Paleolithic period. Later on, i.e., about 26,000 years ago, cracked "boiling stones" were discovered in caves, which have been used by early modern humans. Around 20,000 years ago, pottery has appeared for more conventional boiling. Therefore, for tens of thousands of years, the boiling process has been used for cooking.

The earliest steam engine was the scientific novelties of Hero of Alexandria in the first century CE, called as the aeolipile (https://www.britannica.com/technology/steam-engine [Accessed: December 10, 2023]).This device is the first known one to transform steam energy into a rotary motion. However, like many other early machines, they have demonstrated basic mechanical principles and were simply regarded as a curiosity or a toy and have not been used for any practical purposes.

Only in the seventeenth century, there were attempts to made steam engines for practical purposes (https://www.britannica.com/technology/steam-engine [Accessed: December 10, 2023]). As such, in 1698 Th. Savery patented a pump with hand-operated valves to raise water from mines by suction produced with condensing steam. Somewhere around 1712, Th. Newcomen, has developed a more efficient steam engine with a piston separating the condensing steam from the water. In 1765, J. Watt quite significantly improved the Newcomen engine by adding a separate condenser to avoid heating and cooling the cylinder with each stroke. And finally, he has developed a new engine that rotated a shaft instead of providing the simple up-and-down motion of a pump and added many other improvements to produce a practical power plant.

In 1769, N.J. Cugnot has built the first steam carriage for roads in France (https://www.britannica.com/technology/steam-engine [Accessed: December 10, 2023]). After that R. Trevithick in England was the first to use a steam carriage on a railway; and, in 1803, he built a steam locomotive. In 1829, English engineer G. Stephenson has adapted a steam engine to railways, which became a commercial success. In 1802, W. Symington has built the first practical steamboat. And in 1807, R. Fulton has proposed to use a steam engine for a passenger boat in the United States.

In 1892, L.P. Perkins and W.E. Buck have patented a heat-transmitting device, which was the first two-phase thermosyphon (or, also, it can be named as a wickless heat pipe) operating with boiling-condensation cycle (for more details on these devices, see Bezrodny et al. [1]; Pioro and Pioro [2]).

And only around 1920s, actual research into boiling-heat-transfer phenomena has been started. One of the most significant results, which is important even today, was obtained experimentally by Professor Sh. Nukiyama (http://www.htsj.or.jp/en/nukiyama [Accessed: December 10, 2023]). In 1934, Professor Nukiyama published his pioneering paper entitled "The Maximum and Minimum Values of the Heat Q Transmitted from Metal to Boiling Water under Atmospheric Pressure." In this paper he has clarified and provided an overview of the boiling phenomena in the form of the Nukiyama Curve (nowadays, quite often used just "Boiling Curve") (updated version of this curve is shown in **Figure 1**). Therefore, for more than 100 years, studies into the boiling heat-transfer phenomena at various conditions within a wide range of working fluids, pressures and temperatures, surfaces, etc. are performed non-stop in many research organizations, companies, universities, laboratories, etc. around the world.

2. Classification of boiling cases

In general, boiling is a heat-transfer process during which vapor bubbles are created on a heated surface (nucleate boiling) or inside overheated liquid (bulk boiling) (for properties of various fluids on saturation line, see [3, 4]). Boiling is only possible at subcritical pressures, because at critical and supercritical pressures fluid is a single-phase substance, therefore, there are no such terms as liquid and vapor (see **Figure 2**). However, due to significant variations of all thermophysical properties within critical and pseudocritical regions fluid undergoes a transition from high-density fluid (liquid-like) to low density fluid (vapor-like). Therefore, we have quite similar heat-transfer processes to those at subcritical pressures, which are called pseudo-boiling, pseudo-film-boiling, and deteriorated heat flux (for details, see [5–7]).

An important condition for boiling is that the temperature of a heated surface should be higher, at least, by several degrees than that of saturated liquid (see **Figure 1**). For subcooled boiling this difference in temperatures can be significantly higher.

Boiling process (heat transfer) can be classified in general as the following (for details, see **Figures 1–16**):

1. Nucleate pool boiling (classical case of boiling) (**Figures 1**, 5–7).

2. Transition boiling (quite rare regime) (see **Figure 1**).

3. Film boiling (special boiling regime, which should be usually avoided) (see **Figure 1**).

4. Bulk boiling (the least used type of boiling) (**Figure 9**).

5. Subcooled (**Figure 8**) and saturated boiling (**Figures 3** and **4**).

6. Flow boiling (the most used in industrial applications):

    - Internal boiling (inside tubes, pipes, channels, etc.) (**Figures 6–8**);

    - External boiling (over heated surfaces, annular channels, rod bundles, cross flow, etc.); and

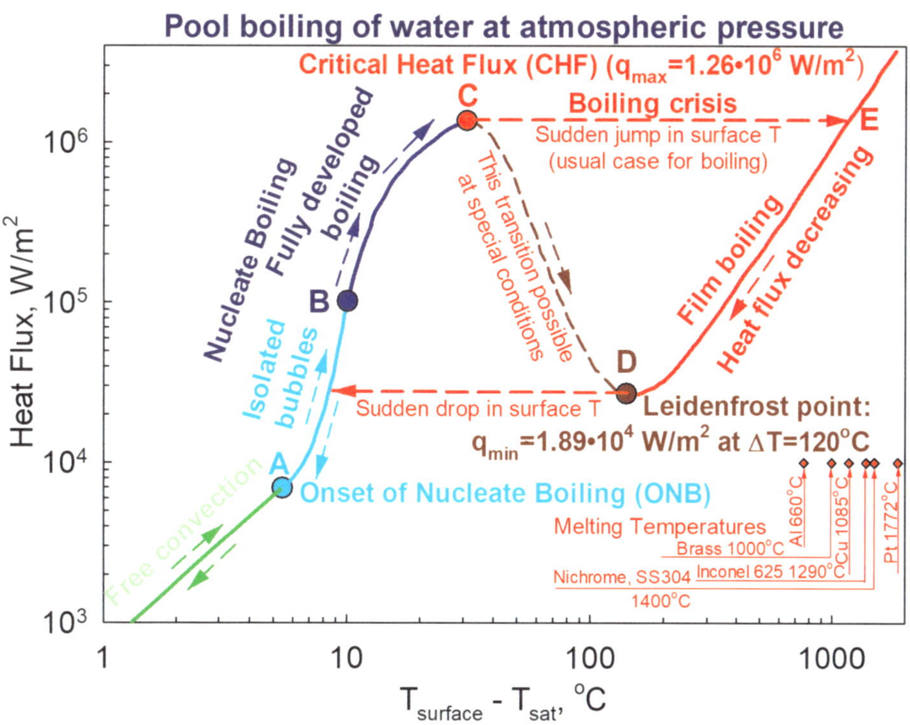

Figure 1.
Boiling curve for saturated water at atmospheric pressure (first time was obtained by Professor Sh. Nukiyama (Tohoku University, Japan) at the beginning of 1930s and for long time called as "Nukiyama's boiling curve"). In the current view the boiling curve is updated with q_{max} *and* q_{min} *values and melting temperatures of selected common metals/alloys.*

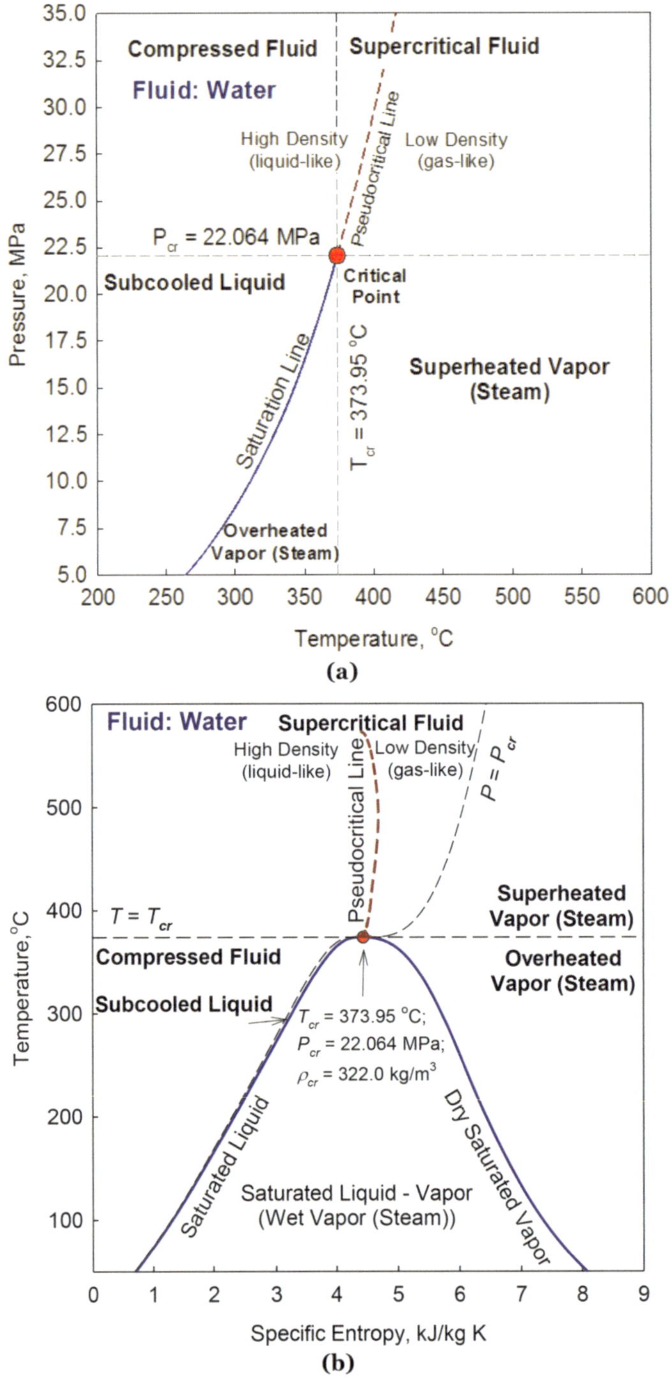

Figure 2.
Thermodynamics diagrams for water: (a) pressure—temperature and (b) temperature—specific entropy.

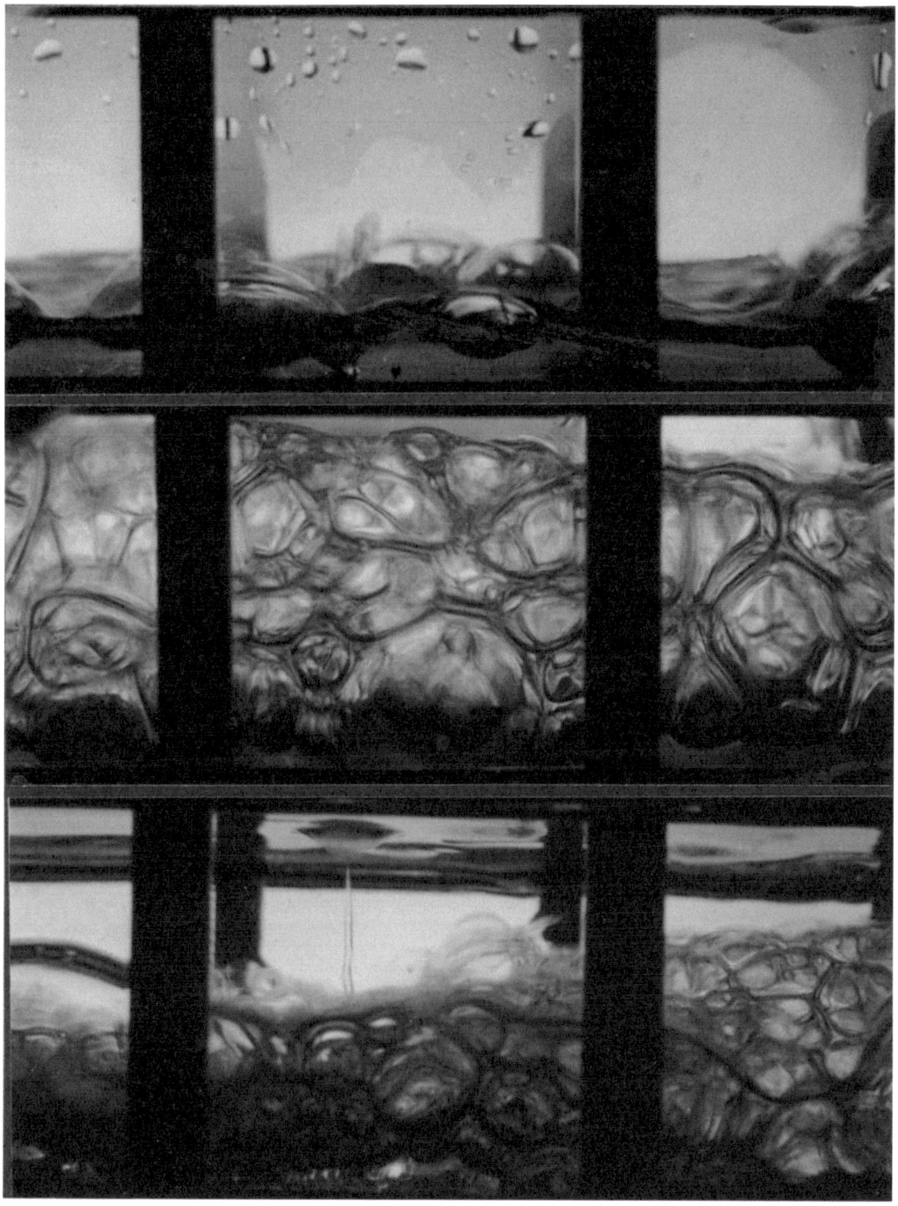

Figure 3.
*Nucleate pool boiling of water on horizontal copper plate (see **Figure** 5 for test section and electron-microscope images of boiling surfaces use in this experimental setup) at sub-atmospheric pressures: Parameters of working fluid —volume 120 ml and level 5.8 mm; height of boiling-condensation chamber 38 mm; scale—pitch between two thread-rods (black vertical posts on photos) equals to 40 mm.*

No. of photo (from top)	Heat flux, kW/m^2	Temperature, °C		Pressure, kPa	HTC, W/m^{2}K
		Wall	Saturation		
1	8.6	56	41	7.8	1343
2	12.2	62	56	16.5	2152
3	29.6	68	58	18.2	3037

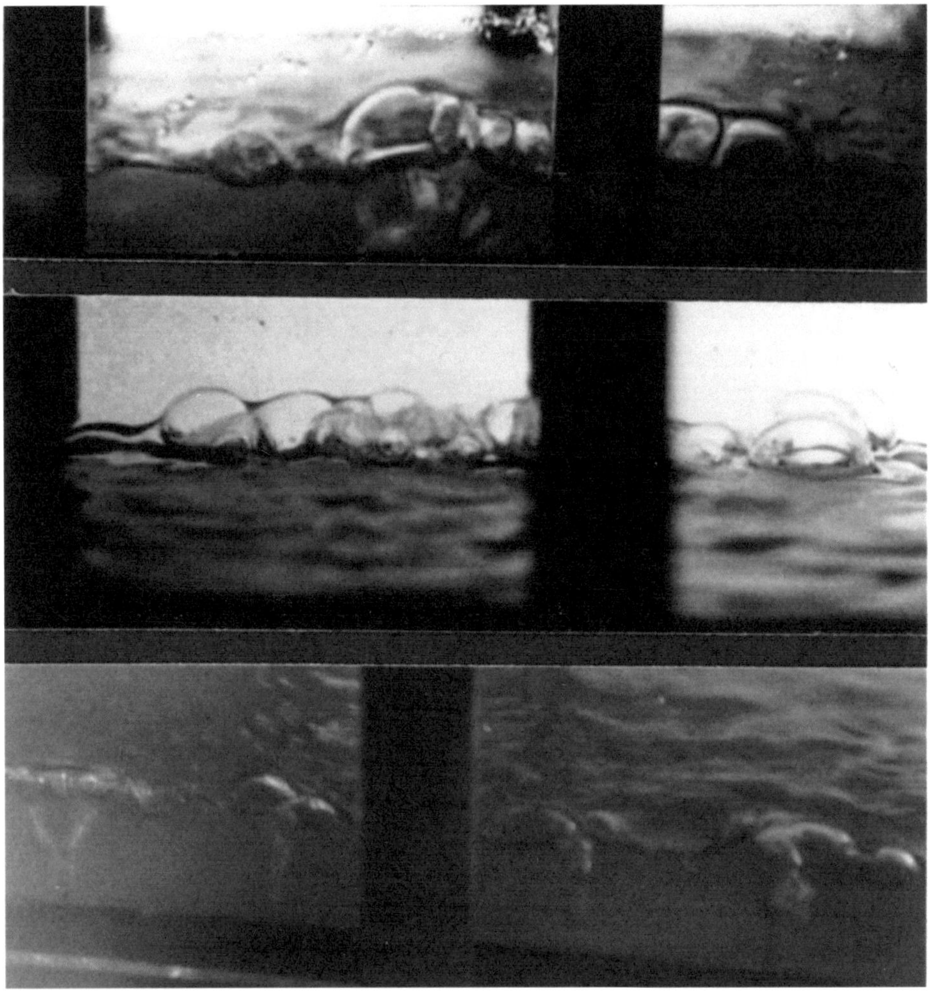

Figure 4.
Nucleate pool boiling of refrigerants on horizontal high-density polyethylene plate (see **Figure** *5 for text section): Parameters of working fluid—volume 120 ml and level 5.8 mm; height of boiling-condensation chamber 38 mm; scale—width of photos equals to 120 mm in actual chamber. Photos 1 and 2—boiling in slots and photo 3 — boiling on surface. Polyethylene has very low surface roughness due to that only several vapor-bubble-generating centers are seen.*

No. of photo (from top)	Working fluid	Heat flux, kW/m^2	Temperature, °C		Pressure, kPa	HTC, W/m^{2}K
			Wall	Saturation		
1	R-113	5.1	32	26	46.6	850
2	R-113	5.1	32	26	46.6	850
3	R-11	10.9	95	30	126.0	168

- Internal or external boiling in or on horizontal (**Figure 16**) or inclined flow geometries.

7. Boiling in two-phase thermosyphons (or wickless heat pipes) (**Figures 9–11**) [1, 2].

8. Boiling in wick structures (in application to heat pipes with wicks) (for details, see [12–14]).

9. Boiling of mixtures (**Figure 12**) (for details, see [15, 16]) and solutions. And,

10. Boiling on enhanced surfaces (fins, ribs, artificial cavities, special surface treatment, etc.) (for details, see [17–19]) or in flow geometries with flow obstructions or turbulizers.

In addition, the following special cases of boiling can be identified:

- Non-isothermal boiling surfaces;
- Variable heat flux along the boiling surface;
- Thin boiling films;
- Rewetting of hot surface;

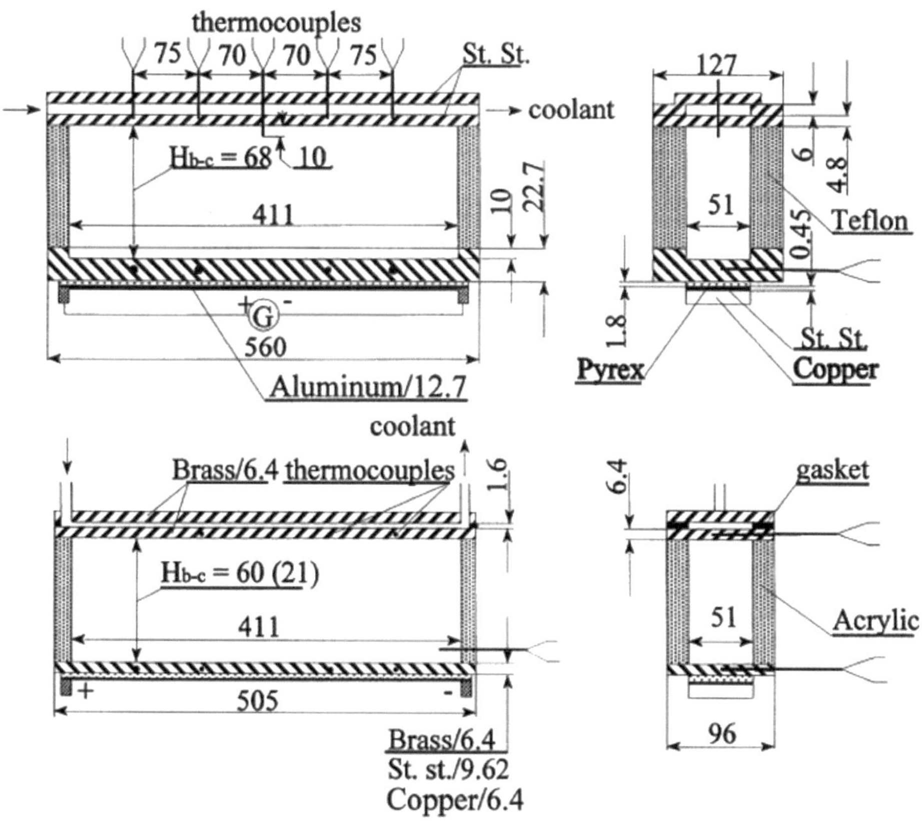

Figure 5.
*Dimensions (in mm) of boiling-condensation chamber: Heating surfaces used—Aluminum, brass, copper, st. st., and high-density polyethylene (see **Figure 15a-e** for electron-microscope images of boiling surfaces and **Tables 3** and **4** for their thermophysical properties and surface-roughness parameters):* H_{b-c}*—height of boiling-condensation chamber.*

- Boiling under conditions of decreased or increased gravitational fields;
- Low pressures (see **Figure 3**);
- Boiling of liquid metals (for details, see [20, 21]);
- Boiling of cryogenic fluids (for details, see [22]);
- Boiling in microchannels; and
- Boiling with nano- /micro-particles.

3. Nucleate pool boiling

Nucleate pool boiling is the first type of boiling to be used by humans and the first one to be investigated. Boiling curve for saturated water at the atmospheric pressure was obtained by Professor Shiro Nukiyama (Tohoku University, Japan) at the beginning of 1930s and for long time was called as the "Nukiyama's boiling curve" (see **Figure 1**). In the current view the boiling curve is updated with $q_{\max}$ and $q_{\min}$ values and melting temperatures of some common metals/alloys. Photos of nucleate pool boiling are shown in **Figures 3** and **4**.

Major nucleate-pool-boiling characteristics (see **Figure 1**) are as the following:

- Onset of nucleate pool boiling (Point A);

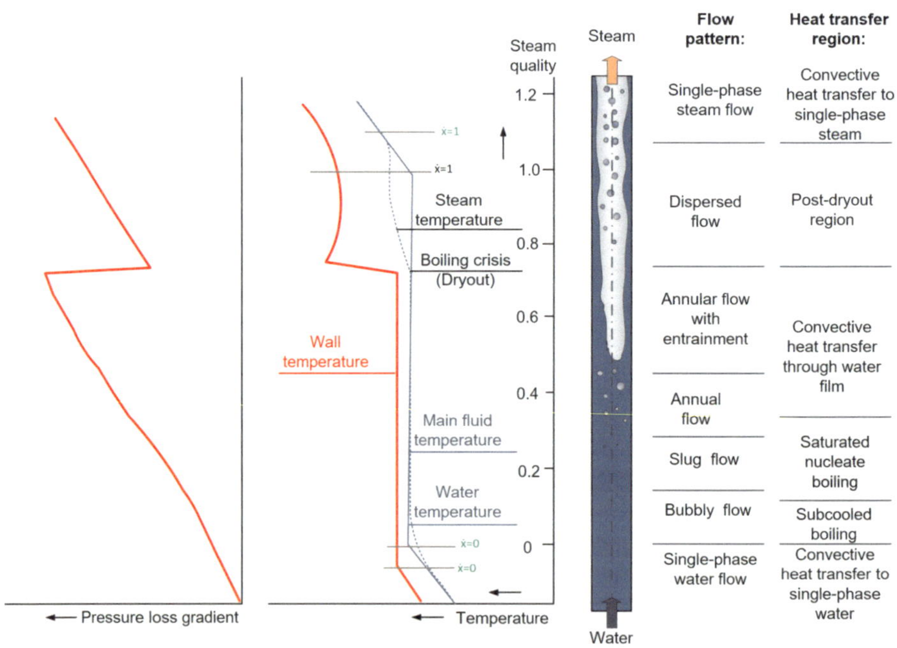

Figure 6.
*Wall- and bulk-fluid-temperature and pressure-loss-gradient profiles in uniformly heated vertical, bare tube at flow boiling (based on **Figure 4** from Siemens: 25JahreBENSONbild_E.doc [Accessed: February 22, 2022]).*

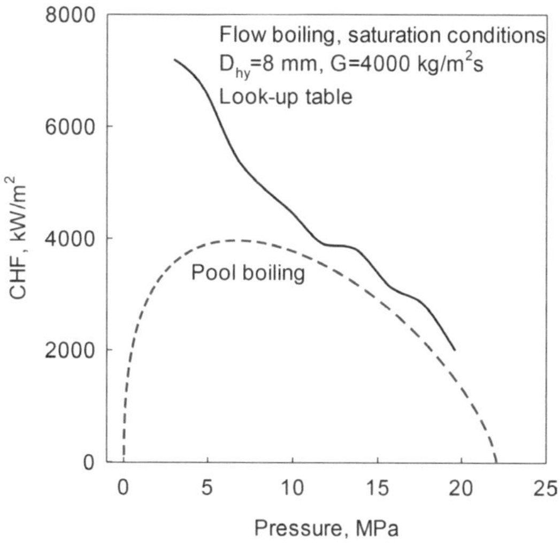

Figure 7.
Critical-heat-flux (CHF) profiles vs. pressure in uniformly heated vertical, bare tube at flow boiling (upper solid curve) and in pool boiling (lower dashed curve).

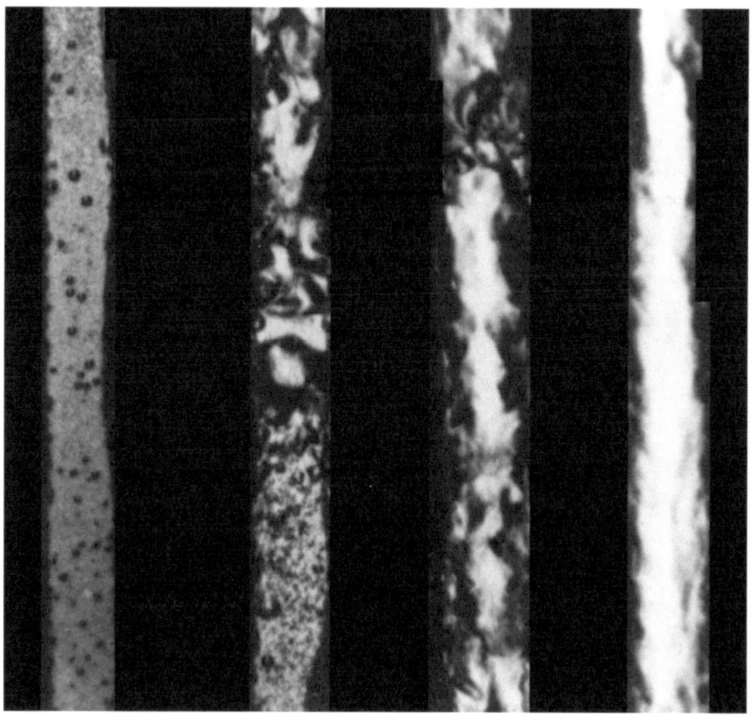

Figure 8.
Flow boiling of water in vertical rectangular channel (8 × 12.5 × 730 mm; two opposite walls—st. st. and other two—transparent acrylic): Pressure 0.1 MPa; inlet velocity 0.036 m/s; subcooling temperature 90°C; heat flux 167 kW/m^2; scale—height of each photo equals to 150 mm in actual test section; width 12.5 mm; from left to right—portions of channel from lower to upper part starting from 130 mm of heated length. Vapor bubbles on photos are shown as black circles. Flow regimes from left bottom to right top (approximately): Bubbly flow; slug flow; annular flow; annular flow with entrainment of droplets; and single-phase steam flow. Liquid film on left and right st. st. walls moves up in photos 2–4.

- Heat Transfer Coefficient (HTC) at nucleate pool boiling (Regions A–B–C);
- Critical heat flux at nucleate pool boiling (Point C);
- HTC at film pool boiling (Region E–D);
- Minimum heat flux at film pool boiling (Point D); and
- Also, in special conditions of controlled experiment we can talk about HTC at transition boiling (Region C–D).

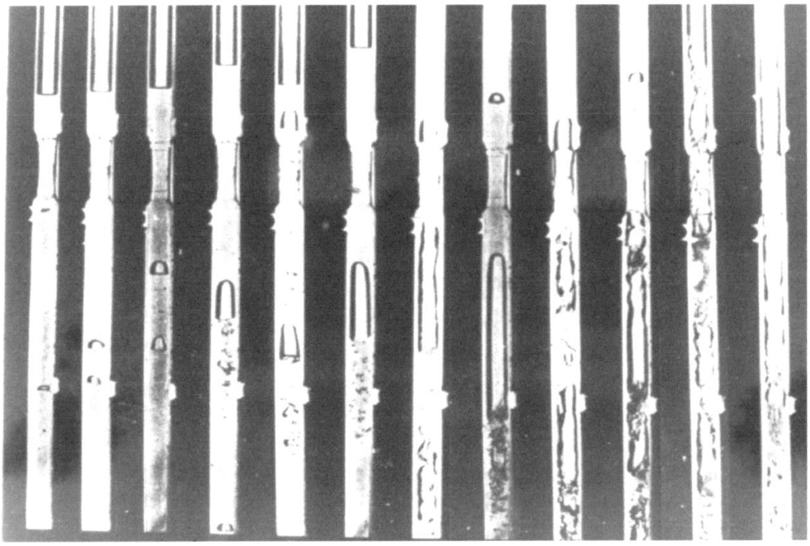

Figure 9.
Bulk boiling in two-phase counter-flow thermosyphon on glass surface: Methylene chloride (R-30), atmospheric pressure, filling charge more than 100% of evaporator volume, evaporator—lower part of thermosyphon and condenser—upper part, in between—short transportation zone), and heat flux increasing from left to right.

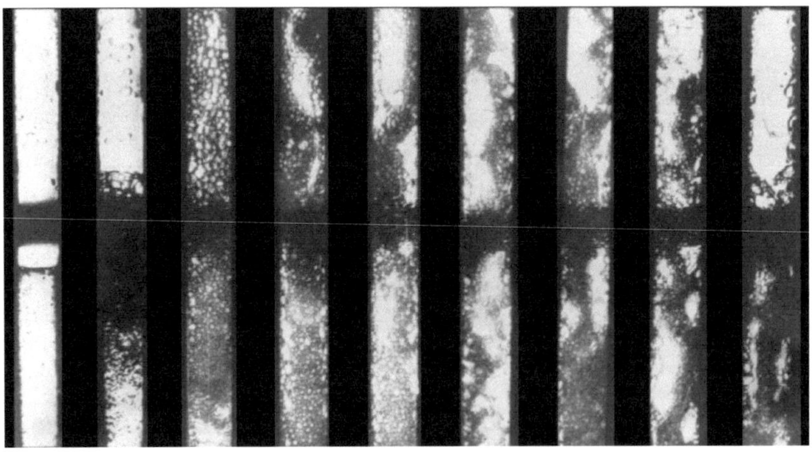

Figure 10.
Nucleate boiling in two-phase counter-flow thermosyphon on st. st. surface: Methylene chloride (R-30), atmospheric pressure, filling charge more 40% of evaporator volume, and heat flux increasing from left to right up to critical heat flux (CHF).

Figure 11.
Nucleate boiling in two-phase counter-flow thermosyphon on metal heated rod (annular-channel evaporator): Water, atmospheric pressure, filling charge 100% of evaporator volume, and heat flux increasing from left to right up to critical heat flux (CHF).

Therefore, for all these Points /Regions we need to have the appropriate correlations (for general correlations, see, for example, Chapter 10 in [9]). In general, there are three internal boiling characteristics such as (for details, see [23–27]; and **Table 1**): (1) vapor-bubble departure diameter, D_b; (2) frequency of vapor-bubbles departure, f; and (3) mean velocity of vapor-bubble growth, $\overline{u}_b = D_b \cdot f_b$. However, these internal-boiling characteristics are not easy to estimate, and their uncertainties

Figure 12.
Nucleate boiling in two-phase counter-flow thermosyphon on metal heated rod (annular-channel evaporator): Water-ethylene-glycol mixture (water boiling temperature ~ 100°C and ethylene-glycol—~200°C), atmospheric pressure, filling charge 100% of evaporator volume, and heat flux increasing from left to right up to critical heat flux (CHF). Photo 1: no boiling—mixture not separated; photos 2–9—mixture is separated, i.e., water (liquid density—958 kg/m^3) boils in the upper part of evaporator and non-boiling ethylene-glycol as liquid (density—993 kg/m^3) transfers heat to boiling water with natural convection in the lower part.

are quite high. Also, there are some theoretical approaches to boiling heat transfer, but, usually, only empirical correlations are used for various nucleate-pool-boiling characteristics/parameters, which are based on well-known and well-defined thermophysical properties. For example, the vapor-bubble departure diameter is usually replaced with [9]:

$$D_b \propto \sqrt{\frac{\sigma}{g\left(\rho_f - \rho_g\right)}}. \tag{1}$$

In 1952, W. Rohsenow has proposed his nucleate pool-boiling correlation, which is the most widely used correlation during the last 70+ years.

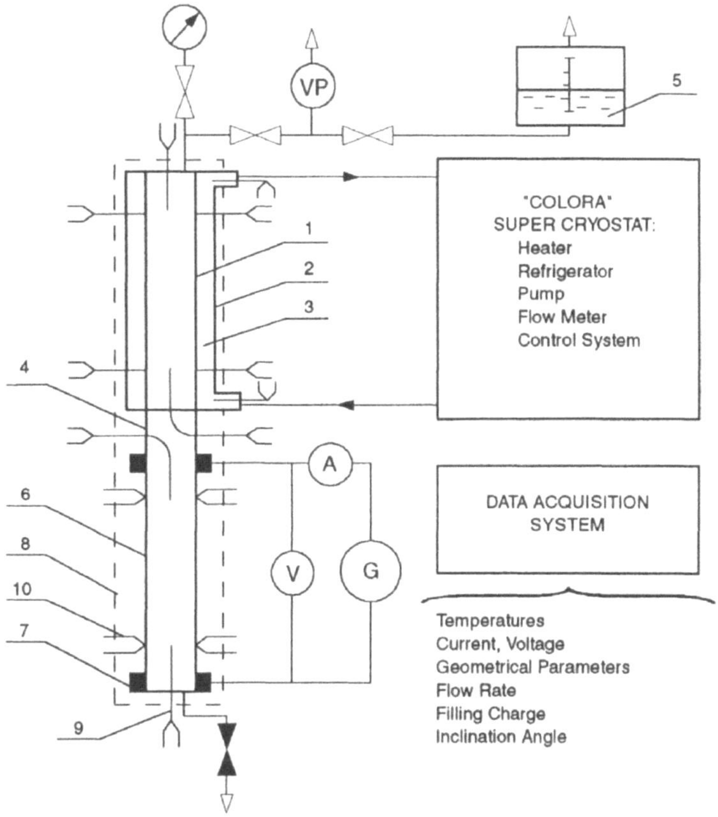

Figure 13.
Universal experimental setup for boiling experiments at various operating conditions and in different flow geometries (current view—test section (to the left) is two-phase counter-flow thermosyphon: 1—condenser; 2—cooling jacket; 3—coolant (antifreezing mixture water-ethylene glycol); 4—transportation (adiabatic) zone; 5—working fluid (WF); 6—evaporator; 7—current terminals; 8—insulation; 9—sheathed thermocouples (fluid/vapor temperatures); 10—wall-temperature thermocouples; A—ammeter; G—electrical generator (power supply); V—voltmeter; and VP—vacuum pump.

The main concept of this correlation is that the heat transfer from the wall directly to the liquid with an increased heat-transfer rate, due to the agitation of liquid by the departing vapor bubbles.

$$\frac{c_{p\ f}\ \Delta T_b}{H_{fg}} = C_{sf}\left[\frac{q}{\mu_f\ H_{fg}}\sqrt{\frac{\sigma}{g\ \left(\rho_f-\rho_g\right)}}\right]^m\left(\frac{c_{p\ f}\ \mu_f}{k_f}\right)^n, \qquad (2)$$

where C_{sf} is constant, depending upon the nature of the heating-surface- fluid combination (see **Table 2**). However, some other well-known correlations do not include any heating-surface parameters/properties or impact of the heating-surface- fluid combination on HTC at boiling (for details, see [10]).

Detailed analysis of the data in **Table 2** has shown that information on the surface-fluid combination is too simplified and, actually, misleading. A thorough analysis of original publications in which C_{sf} values were obtained is presented in the joint

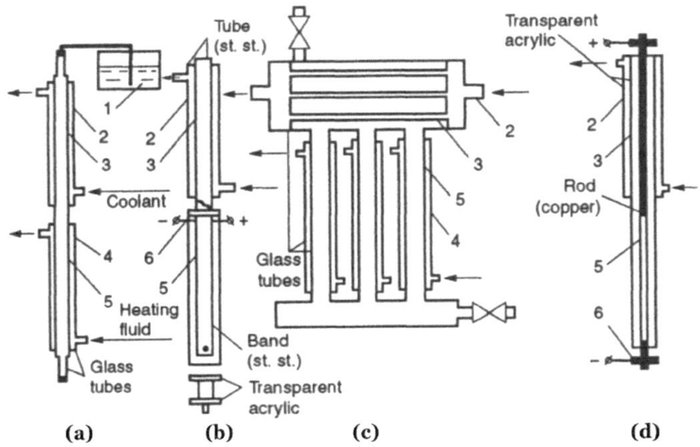

Figure 14.
*Test sections for boiling experiments at various operating conditions: (a) two-phase counter-flow glass thermosyphon (bulk-boiling photos—**Figure 9**); (b) two-phase counter-flow thermosyphon with boiling on st. st. surface (nucleate-boiling photos—**Figure 10**); (c) two-phase counter-flow glass thermosyphon with multiple evaporators and horizontal condenser); and (d) two-phase counter-flow thermosyphon with boiling on st. st. internal tube (annular boiling channel) (nucleate-boiling photos—**Figures 11** and **12**):—fluid-expansion tank; 2—cooling jacket; 3—condenser; 4—heating jacket; 5—evaporator; and 6—current (power) terminals.*

publication by I. Pioro, W. Rohsenow, and S. Doerffer [10, 23] and by Pioro [8] together with the latest Pioro correlation on the pool-boiling heat transfer. This list of C_{sf} values is the most comprehensive and detailed one so far (see Appendix A at the end of this Chapter).

The major problem with correlations, which account for a heating surface-fluid combination, is that these correlations can be used only for a particular heating surface and fluid used in experiments. Otherwise, uncertainties can be very high! On opposite, if correlations, which do not account on a particular heating surface and fluid combination, are used, it is impossible to predict uncertainties of HTCs calculated!

The most important nucleate-pool-boiling characteristic is the Critical Heat Flux (CHF), because if the CHF is reached, the boiling-surface temperature can jump to very high values (beyond 1000°C, see **Figure 1**) and, eventually, the boiling surface can be damaged or even melted. Of course, this temperature rise depends on the type of heating, i.e., for electrical and nuclear heating temperature rise can go far beyond 1000°C. However, if the boiling surface is heated with hot or high-temperature medium, the surface temperature cannot be higher than that of this medium.

The mostly used CHF correlation for pool boiling is as the following (for details, see **Figure 7**):

$$q_{cr} = C_{cr} H_{fg} \left[\sigma g \rho_g^2 \left(\rho_f - \rho_g \right) \right]^{0.25}, \tag{3}$$

where C_{cr} is constant with the average value of 0.15. However, in reality, this constant can be within the range of 0.08–0.28! This correlation was obtained through a dimensional analysis by S.S. Kutateladze in Russia in 1948 and through a hydrodynamic-stability analysis by N. Zuber in the United States in 1958 [9].

Therefore, in conclusion we can say that in spite of more than 100 years of active research into the pool-boiling phenomena, we have failed to develop universal HTC,

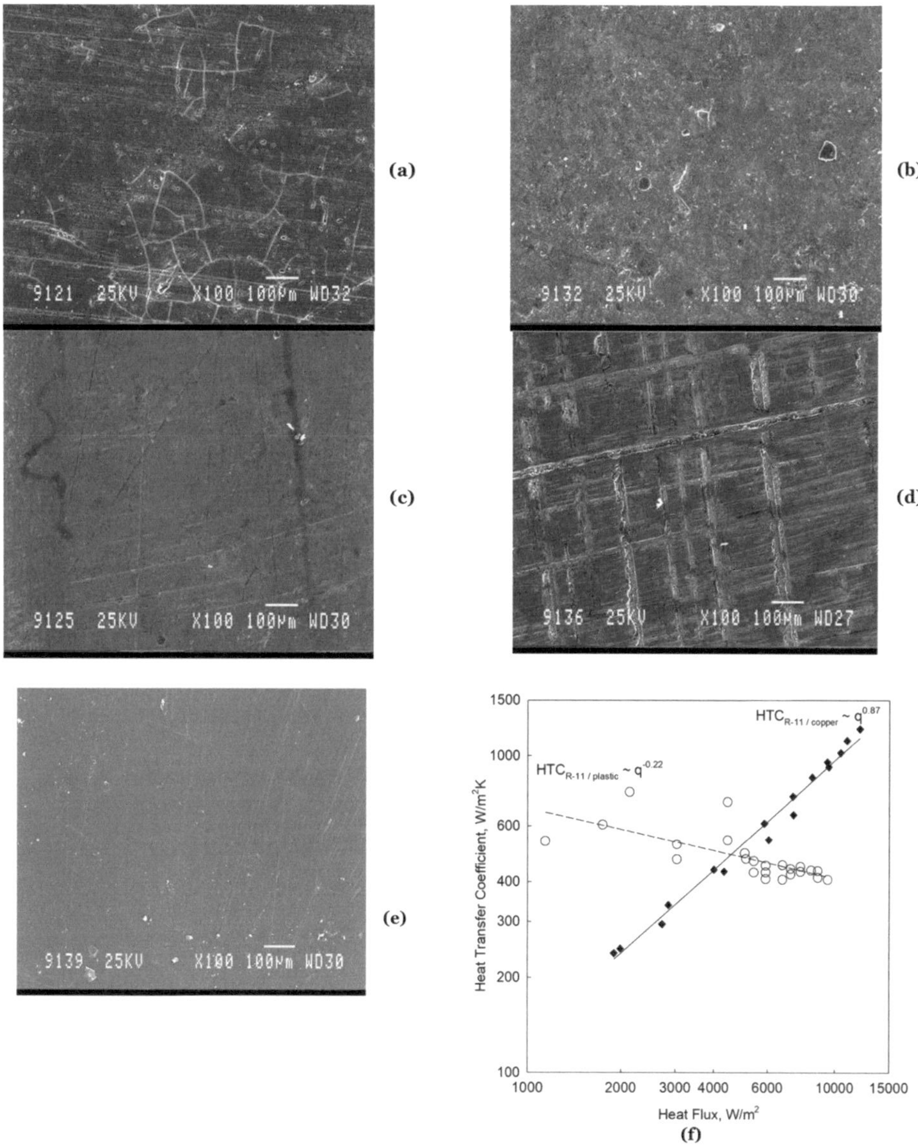

Figure 15.
a-e. Electron-microscope images (enlargement × 100) of plates made of: (a) aluminum—R_q = 4.5 µm, R_{sk} = 0.47 µm; (b) copper—R_q = 1.7 µm, R_{sk} = 0.38 µm; (c) brass—R_q = 0.7 µm, R_{sk} = −1.3 µm; (d) st. st.—R_q = 0.6 µm, R_{sk} = 0.19 µm; and (e) polyethylene high density. For thermophysical properties and surface-roughness parameters, see ***Tables 3*** *and* ***4****, respectively, and Appendix A). Details on experimental setup with these heated plates and experimental data are presented in [8]. f. Effect of heat flux on HTC at nucleate pool boiling of R-11 on copper (R_q = 1.7 µm) (b) and plastic (PHD) (e) large-size plates.*

CHF, minimum heat flux, film boiling, and other correlations with a reasonable accuracy, which can be applied to various heating surfaces with different thermophysical properties, surface-roughness parameters and microgeometry, wall thickness, orientation in space, and different boiling fluids within a wide range of operating conditions!

The thermophysical properties of boiling surfaces are listed in **Table 3**.

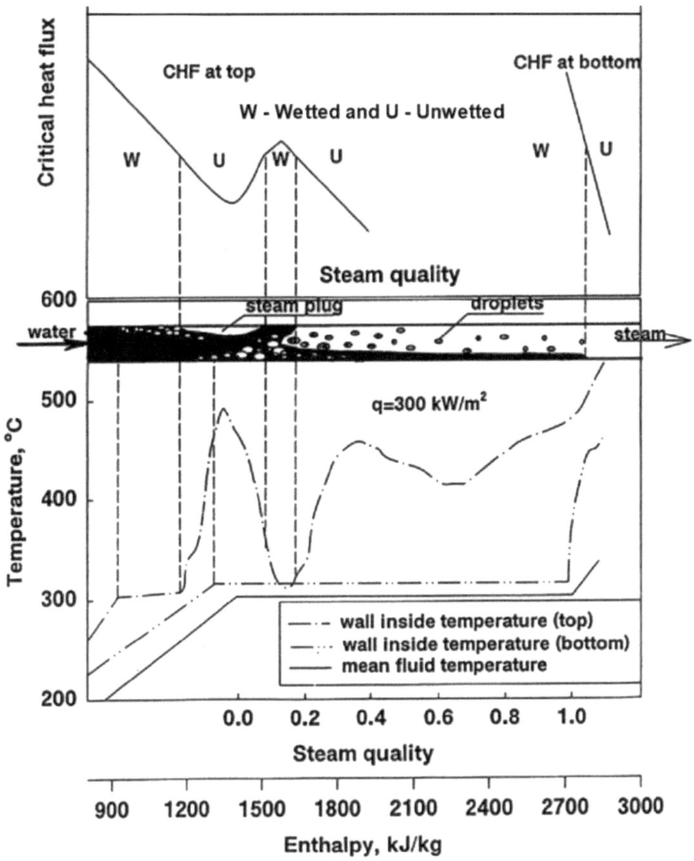

Figure 16.
Effect of gravity on boiling heat transfer and CHF in horizontal tube: Water, P = *10 MPa,* G = *500 kg/m²s,* D = *24.3 mm (based on Kohler and Hein [11]) (Courtesy of NRC, USA).*

Fluid	P_{sat}, MPa	P/P_{cr}·10^3	Boiling surface	D_b, mm	f_b, s^{-1}	u_b, mm/s
Water	0.1	4.52	Permalloy	2.5	61	153
			Brass	2.3	67	157
			Copper	2.8	56	157
Freon-12	0.1	23.6	Permalloy	0.7	84	59
			Brass	0.7	99	69
			Copper	0.7	91	64
CCl_4	0.1	22.0	Permalloy	1.1	110	121
			Brass	1.1	108	119
			Copper	1.1	106	117
Ethanol (96.5%)	0.1	15.6	Permalloy	1.0	114	114
			Brass	1.1	112	123
			Copper	1.2	98	118
Methanol	0.1	12.6	Permalloy	1.7	74	124

Fluid	P_{sat}, MPa	$P/P_{cr}\cdot 10^3$	Boiling surface	D_b, mm	f_b, s^{-1}	u_b, mm/s
n-Butanol	0.1	20.2	Permalloy	1.1	106	111
Benzene	0.1	20.3	Permalloy	1.0	99	99

Table 1.
Internal boiling characteristics of various fluid-surface combinations [23, 24].

No.	Surface-fluid combination	C_{sf}	n
1	Water-copper Scored Polished	0.0068 0.0128	1.0 1.0
2	Water-stainless steel Chemically etched Mechanically polished Ground and polished	0.0133 0.0132 0.0080	1.0 1.0 1.0
3	Water-brass	0.0060	1.0
4	Water-nickel	0.0060	1.0
5	Water-platinum	0.0130	1.0
6	n-Pentane-copper Polished Lapped	0.0154 0.0049	1.7 1.7
7	Benzene-chromium	0.0101	1.7
8	Ethyl alcohol-chromium	0.0027	1.7

Table 2.
Values of C_{sf} for various surface-fluid combinations [9].

No.	Boling surface material	k W/m K	c J/kg K	ρ kg/m^3
1	Copper	401	385	8933
2	Aluminum	237	903	2702
3	Brass	110	380	8530
4	St. St.	145	477	7900
5	Polyethylene High Density (PHD)	0.35–0.49	1330–2400	955–961

Table 3.
Thermophysical properties of boiling surfaces (extended plates) at 27°C (listed according to decreasing thermal-conductivity values) [9].

A laser profilometer was used to determine the surface-roughness parameters that are listed in **Table 4**. The characteristics of the laser profilometer itself were as follows:

Vertical measuring range	± 300 μm.
Scanning speed	80 mm/min
Number of measured points	2000
Wavelength cut-off (L_c)	1.0 mm
Stylus filter (N_f)	50 μm
Scan length	10 mm

Plate material	R_a	R_q	R_p	R_v	R_t	R_{pm}	R_z	R_{3z}	HSC	S_m	λ_a	λ_q	Δ_a	R_{sk}
	μm	**μm**	**μm**	**μm**	**μm**	**μm**	**μm**	**μm**	**—**	**μm**	**μm**	**μm**	**deg.**	**μm**
Al	**3.6**	**4.5**	14	14	28	12	22	17	89	89	88	82.9	0.26	0.47
Copper	**1.4**	**1.7**	7.2	4.4	12	5.2	9	6.4	68	117	109	107	0.08	0.38
Brass	**0.5**	**0.7**	2.4	5.1	7.4	1.7	3.9	2.4	126	63	65	70.1	0.05	−1.3
St. St.	**0.5**	**0.6**	3.4	2.5	5.9	2.1	3.8	2.5	123	64	65	65.2	0.05	0.19
	Minimum and maximum values of roughness parameters													
Min	0.5	0.6	2.4	2.5	5.9	1.7	3.8	2.4	68	63	65	65.2	0.05	−1.3
Max	1.4	4.5	14	14	28	12	22	17	126	117	109	107	0.26	0.47

Table 4.
Average surface-roughness parameters of boiling surfaces (extended plates) (listed according to decreasing surface-roughness (R_q/R_a) values) (for descriptions of all surface roughness parameters, see below or in [10]).

Explanations to **Table 4**.

Simple-Roughness-Amplitude Parameters.

Mean parameters.

R_a average roughness: area between the roughness profile and its mean line or its integral of the absolute value of the roughness-profile height over the evaluation length. The average roughness is the most commonly used parameter in surface-finish measurements.

R_q root-mean-square roughness (rms roughness): average roughness parameter calculated as a square root from another integral of the surface-roughness profile. Root-mean-square roughness was a commonly used parameter in the past; however, nowadays it has been replaced with R_a in metal-machining specifications. Usually (but not necessarily), R_q is 1.1–1.3 times larger than R_a.

Extreme parameters.

R_p peak roughness (height of the highest peak in the roughness profile over the evaluation length).

R_v depth roughness (depth of the deepest valley in the roughness profile over the evaluation length).

R_t total roughness (vertical distance from the deepest valley to the highest peak), $R_t = R_p + R_v$.

Mean-extreme parameters.

R_{pm} mean-peak roughness (average peak roughness over the sample length).

R_z mean-total roughness (average value of the five highest peaks plus the five deepest valleys over the evaluation length).

R_{z3} mean-total roughness of third extremes parameters (average vertical distance from the third deepest valley to the third highest peak).

Mean-extreme parameters are less sensitive to single unusual features, such as artificial scratches, gouges, burrs, etc.

Roughness-spacing parameters.

HPC High-Peak Count per length (number of peaks per length that cross above a certain threshold and then go back below it).

Mean-roughness-spacing parameters.

S_m mean spacing between peaks (peaks cross above a mean line and then go back below it).

λ_a average wavelength of surface.

λ_q rms (root-mean-square) average wavelength of surface.

Roughness-hybrid parameters.

Δ_a average of absolute slope of roughness profile over the evaluation length.

L_o actual profile length (in all measurements, this was 8 mm).

Statistical parameters.

R_{sk} skewness (this parameter represents the profile variation symmetry over its mean line). Surfaces with $R_{sk} < 0$ have fairly deep valleys in a smoother plateau. Surfaces with $R_{sk} > 0$ have fairly high spikes, which protrude above a flatter average.

4. Flow boiling

Flow boiling is boiling with forced convection, which is the most used type of boiling in industry [28], especially, in thermal- and nuclear-power industry [29, 30] and air-conditioning and refrigeration industry [31, 32]. In the thermal-power industry gas-fired and coal-fired (or fossil-fuel-fired) power plants are used, many of which equipped with the subcritical-pressure Rankine steam-turbine cycle (see **Figure 17a**) [33, 34]. In nuclear-power industry of the world there are 441 nuclear-power reactors connected to electrical grids of which 60 reactors are Boiling Water Reactors (BWRs) including several Advanced BWRs (ABWRs). Moreover, all current nuclear-power reactors are connected only to Rankine steam-turbine power cycles in which boiling takes place in steam generators (in BWRs and ABWRs saturated steam is generated inside reactors (for details, see **Figures 17b**, **18**, and **19**).

More information on all current and future nuclear-power reactors and their power cycles can be found in [29].

The main advantage of using flow or forced-convection boiling is very high HTCs compared to other types of heat transfer (see **Table 5**).

Major flow-boiling characteristics (see **Figure 6**) are as the following:

- Onset of subcooled nucleate boiling;
- Onset of significant void;
- Heat transfer at flow boiling;
- CHF at flow boiling (for details, see **Figure** 7); and
- Post-DryOut (PDO) heat transfer.

In general, these flow-boiling characteristics are quite similar to those of pool boiling. It is impossible to provide correlations for all cases of pool boiling as well as of flow boiling. However, this Chapter contains a list of references and bibliography, which have quite a large number of various cases covered and correlations provided.

Figures 20–23 show specifics of flow boiling in circular tubes, and the experimental setup for these experiments is shown in **Figure 24**. This study covers only two fluids: water and R-134a. To enable a comparison of CHF results between water and R-134a, the R-134a results were converted to their water-equivalent values using the following CHF fluid-to-fluid modeling relationships. It has been shown for vertical

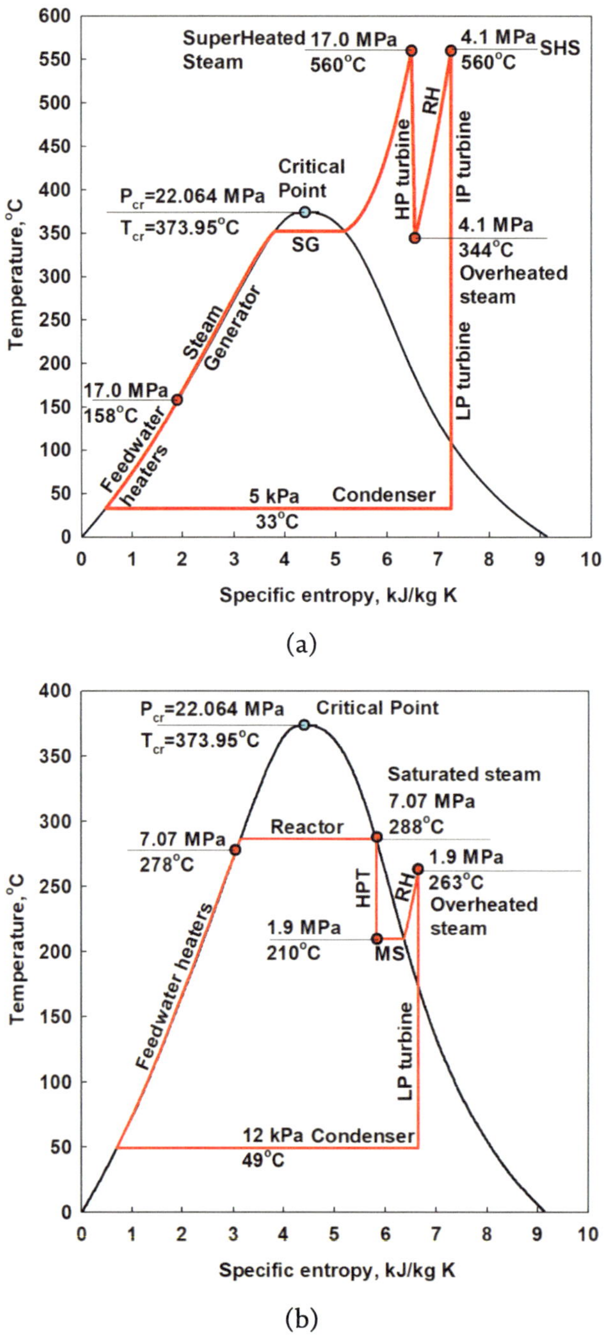

Figure 17.
Temperature (T) vs. specific entropy (s) simplified diagram of subcritical-pressure Rankine power cycle: (a) fossil-fuel-fired thermal power plant with superheated primary- and secondary-steam reheat and (b) advanced boiling water reactor (ABWR) with saturated primary steam and overheated secondary-steam (see ***Figure 19****). Abbreviations: HPT—high-pressure turbine; LP—low pressure; MS—moisture separator; RH—re heater; SG—steam generator; and SHS—super heated steam.*

Figure 18.
Heat-transfer tubes are installed into steam generator (SG) of PWR (1110 MW_{el} and 3060 MW_{th}) (in total four SGs per one reactor) (courtesy of Rosatom): https://www.flickr.com/photos/rosatom/36999718643/in/album-72157675727427445/ [Accessed: December 10, 2023]; Photo by E. Lyadov, Atommash, 2016.

Primary reactor-coolant pressure, MPa (T_{sat} = 348.4°C)	16.2
Coolant temperature at reactor inlet, °C	298
Coolant temperature at reactor outlet, °C	329
Steam-generator pressure, MPa (T_{sat} = 285.8 °C)	7.0

Reactor coolant is inside tubes and Rankine-cycle feedwater heated and boiling outside tubes.

tubes (see in [36]) that if the fluid-to-fluid modeling relationships are satisfied, i.e., $L_R = L_W$, $D_R = D_W$ (geometric similarity),

$$\left[\frac{G}{\left(\frac{\rho_f \sigma}{D}\right)^{0.5}}\right]_R = \left[\frac{G}{\left(\frac{\rho_f \sigma}{D}\right)^{0.5}}\right]_W \text{ and } \left(\frac{\rho_f}{\rho_g}\right)_R = \left(\frac{\rho_f}{\rho_g}\right)_W \text{ (hydrodynamic similarity),}$$

$x_{cr\,R} = x_{cr\,W}$ (thermodynamic similarity), then the dimensionless CHF expressed as $\left[\frac{q_{cr}}{G h_{fg}}\right]_R = \left[\frac{q_{cr}}{G h_{fg}}\right]_W$ will also be the same for both fluids. Even though the study deals with experiments in R-134a, the water CHF look-up table is also used as a reference, as this table represents an already normalized CHF database for water.

The look-up table data were normalized to tubes with an 8 mm ID; to obtain the CHF for a different diameter, a simple correction can be applied: $\frac{CHF_D}{CHF_{D=8\ mm}} = \left(\frac{D}{8}\right)^{-0.5}$, where D is the Inside Diameter (ID) of a circular tube in mm, D = 8 mm is the reference tube ID.

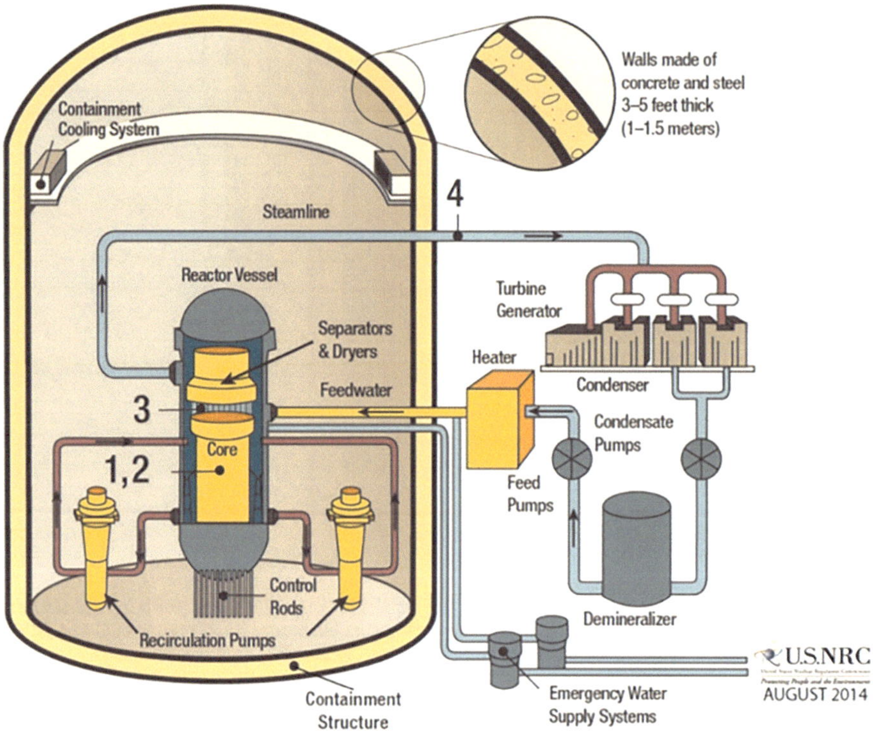

Figure 19.
Simplified layout of typical boiling water reactor (BWR) NPP (courtesy of U.S. NRC): General basic features—(1) thermal neutron spectrum; (2) uranium-dioxide (UO_2) fuel; (3) fuel enrichment about 3%; (4) direct cycle with steam separator (steam generator and pressurizer are eliminated), i.e., single-flow circuit (single loop); (5) reactor pressure vessel (RPV) with vertical fuel rods (elements) assembled in bundle strings cooled with upward flow of light water (water and water-steam mixture); (6) reactor coolant, moderator and power-cycle working fluid are the same fluid; (7) reactor coolant outlet parameters: Pressure about 7 MPa and saturation temperature at this pressure is about 286°C; and (8) power cycle—subcritical-pressure regenerative Rankine steam-turbine cycle with secondary-steam reheat (for details, see ***Figure 17b****). The largest BWR has installed capacities: 1435 MW_{el} and ~ 4500 MW_{th}.*

No.	Coolant	Heat Transfer Coefficient, kW/m^2K
1	Na (forced convection)	55–85
2	**Boiling water (flow boiling)**	**60**
3	CANDU reactor	50
4	Pb (forced convection)	25–35
5	Water (single-phase forced convection)	30
6	Pb-Bi (forced convection)	20–30
7	SuperCritical Water (SCW)	10–25
8	He (rough surface)	10
9	CO_2 (high pressure)	2–5

Table 5.
Selected typical heat-transfer-coefficient (HTC) ranges of various coolants [35].

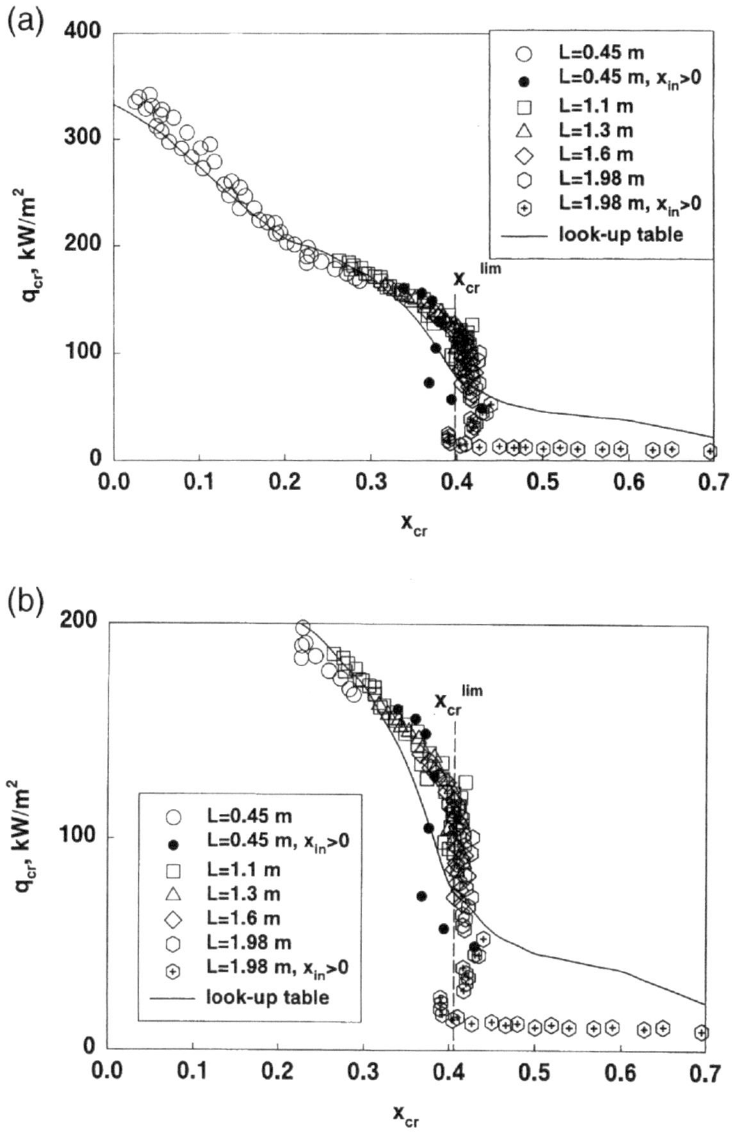

Figure 20.
CHF vs. critical quality at flow boiling in vertical bare circular tube (ID 6.92 mm, OD 7.93 mm, heated length 0.45–1.98 m, material Inconel): R-134a, P = 1.67 MPa, and G = 1000 kg/m²s. (a) Full scale and (b) the same as in (a), but in enlarge scale. (For details, see [36]).

The largest by scale and the most expensive experiments are performed in nuclear-power industry to determine the abovementioned flow-boiling characteristics, because any new bundle design or even updated one requires the exact knowledge of these characteristics. Samples of several bundle-string designs and fuel channel are shown in **Figures 25** and **26**, respectively. Also, in nuclear reactors usually axial and radial heat fluxes are not uniform. These specifics increase significantly the complexity of manufacturing test sections/stations (directly-heated with electrical current thin-wall tubes have to be with variable wall thicknesses) (see **Figures 27** and **28**).

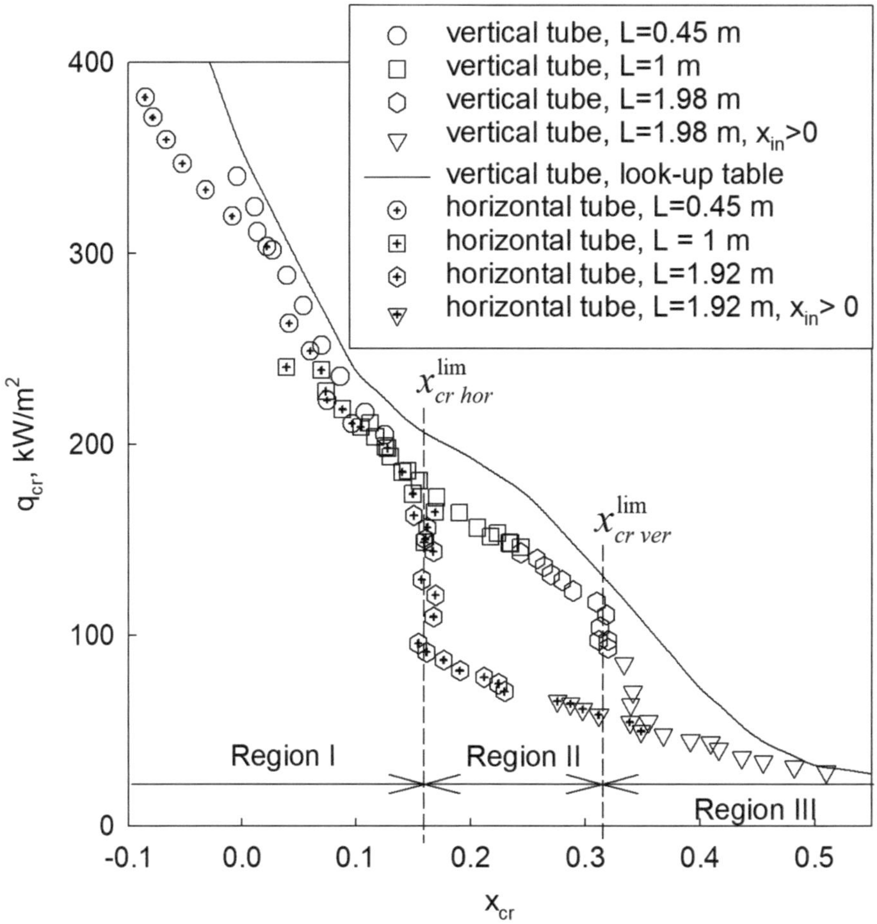

Figure 21.
CHF vs. critical quality at flow boiling in vertical and horizontal circular tubes (ID 6.92 mm, OD 7.93 mm, heated length 0.45–1.98 m, material Inconel): R-134a, P = *1.31 MPa, and* G = *2000 kg/m²s. (For details, see [37]).*

All experiments with bundles are performed with electrically-heated bundle strings, so-called, bundle simulators (for details, see [39, 40]). Therefore, such bundle strings are usually made of Inconel or stainless steel thin-wall tubes and can cost millions of dollars. Also, experimental setups are very sophisticated in terms of measuring devices and require quite large power supplies, e.g., for water experiments with the full-scale bundle string as shown in **Figure 26**, it can be up to 15 MW_{el}, but if modeling fluid (usually, R-134a) is used for additional set of experiments, power requirement can be significantly lower, i.e., up to 1.8 MW_{el} (**Figures 29** and **30**).

To be able to scale operating conditions in water into those of R-134a and vice versa to scale PDO results from R-134a into water data the following scaling laws have been used:

For pressure:

$$\left(\frac{\rho_f}{\rho_g}\right)_R = \left(\frac{\rho_f}{\rho_g}\right)_W \tag{4}$$

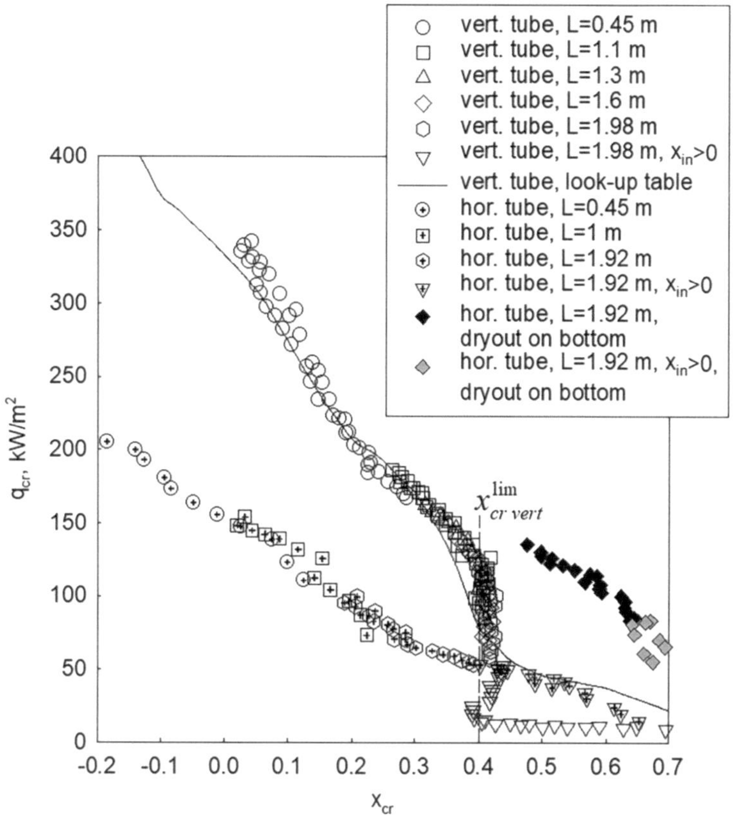

Figure 22.
CHF vs. critical quality at flow boiling in vertical and horizontal tubes (ID 6.92 mm): R-134a, P = 1.67 MPa *and G = 500 kg/m²s.*

For mass flux:

$$\left[\frac{G\ D_{hy}}{\mu_g}\right]_R = \left[\frac{G\ D_{hy}}{\mu_g}\right]_W \tag{5}$$

For PDO HTC:

$$\left(\frac{h_{PDO}\ D_{hy}}{k_g}\right)_R = \left(\frac{h_{PDO}\ D_{hy}}{k_g}\right)_W;\text{where}\quad h_{PDO} = \frac{q}{T_w - T_{sat}} \tag{6}$$

$x_R = x_W$, where x is the thermodynamic quality.
Dimensionless CHF expressed as:

$$\left[\frac{q_{cr}}{G h_{fg}}\right]_R = \left[\frac{q_{cr}}{G h_{fg}}\right]_W \tag{7}$$

It should be noted that the most important parameters for BWR/ABWR bundle-string experiments are HTC, CHF, and PDO heat transfer. Moreover, even for PWRs (the largest PWR is the EPR (Evolutionary Power Reactor) by former company Areva, currently, by EDF (France): 1670 MW_{el} and ~ 4590 MW_{th}) and PHWRs (largest

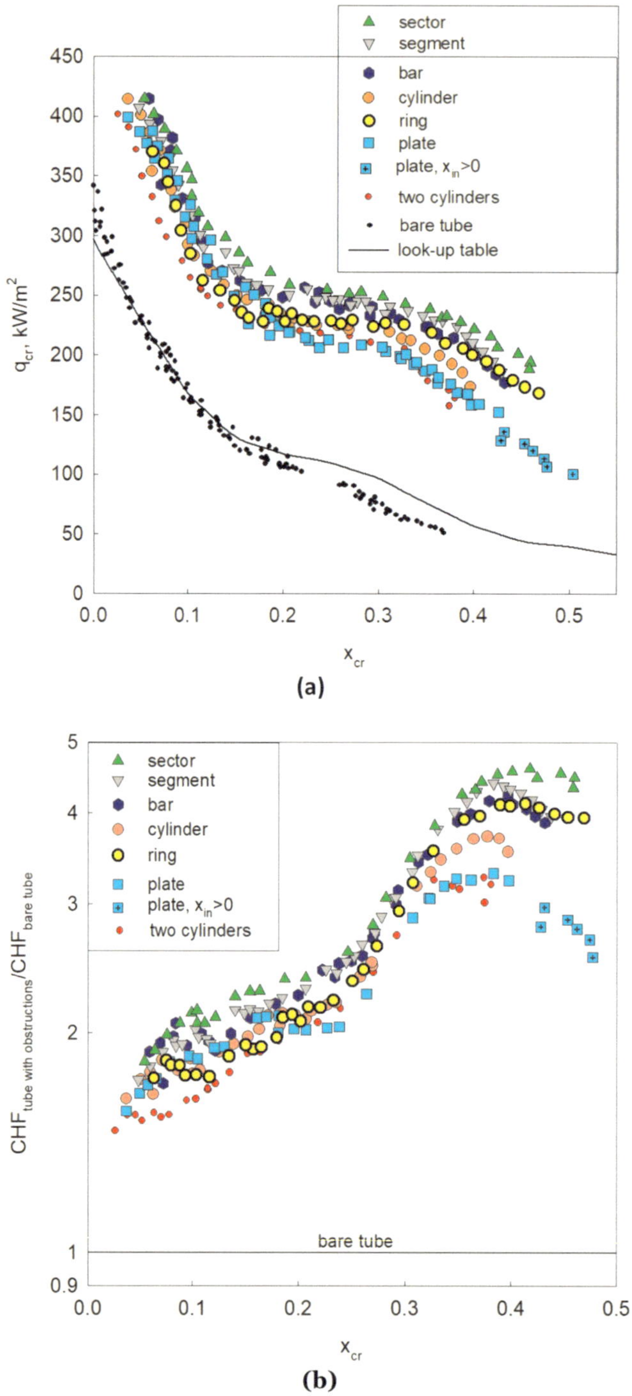

Figure 23.
(a) CHF vs. critical quality and (b) CHF enhancement vs. critical quality—Flow boiling in vertical circular tubes (ID 6.92 mm, OD 7.93 mm, heated length 0.45–2 m, material Inconel) without flow obstructions (i.e., bare) and with various flow obstructions: R-134a, P = *1.67 MPa, and* G = *3000 kg/m²s. (For details, see [38]).*

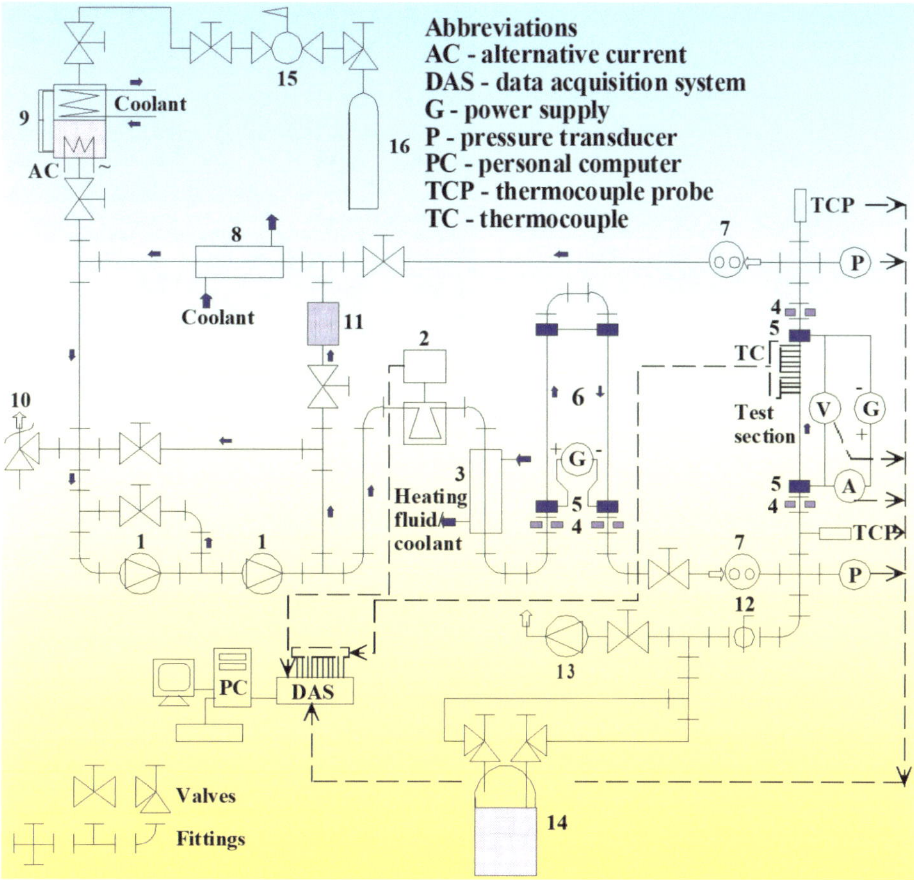

Figure 24.
Experimental thermalhydraulics R-134a loop: 1—gear pump; 2—coriolis-type mass flow meter; 3—preheater; 4—dielectric fittings, 5—Current (power) terminals; 6—electrical preheater; 7—sight glass; 8—condenser; 9—pressurizer; 10—pressure-relief valve; 11—refrigerant filter-dryer; 12—ball valve; 13—vacuum pump; 14—refrigerant storage tank; 15—pressure reducer; and 16—N_2 container.

PHWR is the CANDU-9 reactor (CANada Deuterium-Uranium)) by AECL (Canada): 878 MW_{el} and $\sim$ 2750 MW_{th}), which are not cooled with boiling light or heavy water, CHF and PDO at flow boiling should still be determined.

Figure 31 shows the surface-temperature map for Element 35 at a pressure of 0.98 MPa and mass-flow rate of 9.6 $kg{\cdot}s^{-1}$ with 28% overpower[1]. At high overpowers, dry patches coalesced at some angular locations and the maximum axial dry patch approached the complete length of the element. A full-length axial dry patch on an element could not be measured due to the limited traveling distance of the thermocouple drive unit. Based on the variation of surface temperature with axial distance, the full-element dryout was achieved at several high-power levels.

Figure 32 shows a new application for boiling process such as an ultimate emergency cooling of the molten nuclear-reactor core (corium) during a severe nuclear

[1] Overpower is defined as: channel-power / critical-power

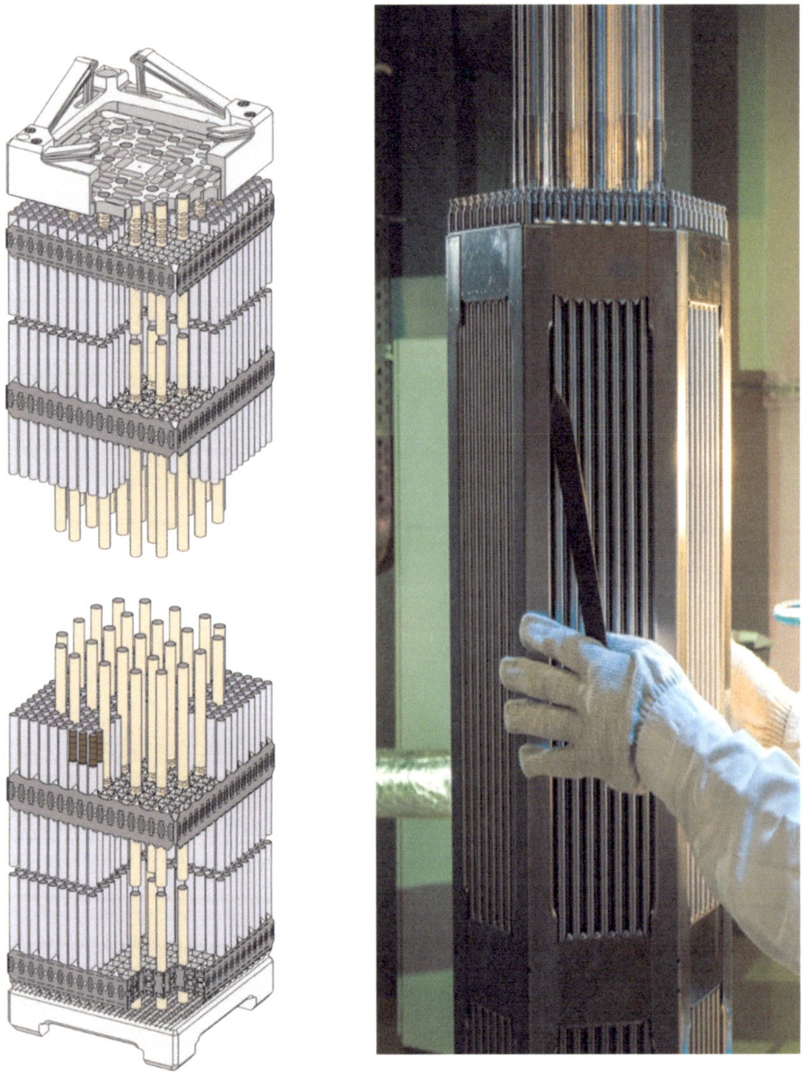

(a) (b)

Figure 25.
Designs of fuel-bundle strings or assemblies of two pressurized water reactors (PWRs): (a) square cross section (courtesy & copyright by MHI) and (b) hexahedron cross section (courtesy of ROSATOM) (Photo by A. Antonov, 2015): https://www.flickr.com/photos/rosatom/25761756447/in/album-72157692396689951/ [Accessed: December 10, 2023].

accident (modern feature for Generation-III^{+} reactor designs) (for details, see [29]). This new safety feature is in response to the Chernobyl NNP severe accident (April of 1986), when a large pressure-channel reactor (RBMK-1000: 1000 MW_{el} and $\sim$ 3200 MW_{th}) was completely melted, and there was a possibility for a corium to damage a concrete foundation of the reactor).

This new application of the boiling process has started to be implemented in new reactor's designs, but this unusual type of boiling is well-known for the Mother nature

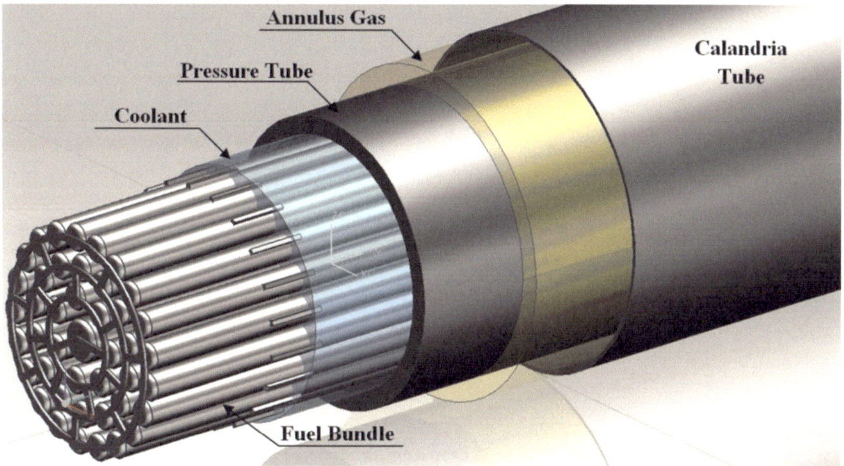

Figure 26.
3-D image of pressurized heavy-water reactor (PHWR) fuel channel with 43-element bundle (based on AECL design; prepared by Dr. W. Peiman).

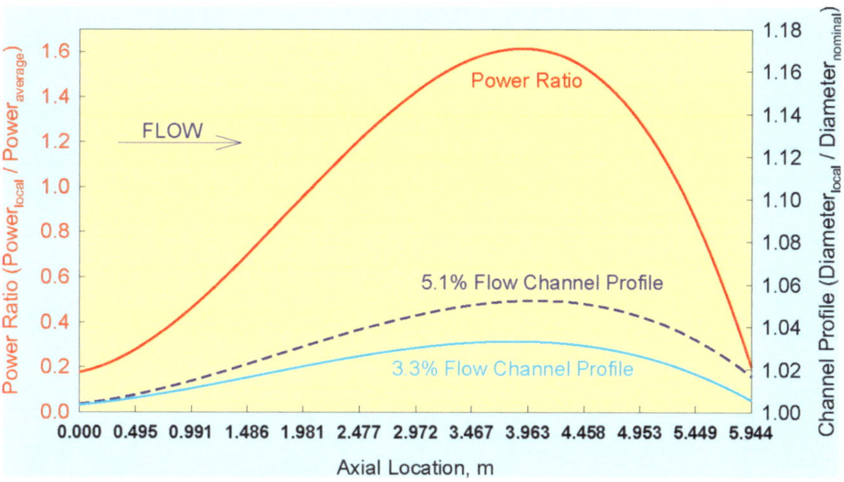

Figure 27.
Sample of stylized axial power profile (APP) or axial heat flux profile (AHFP) used for critical heat flux (CHF) tests and pressure-tube creep profiles: 3.3% for ~10–15 years of operation and 5.1% for ~20–30 years of operation (based on report COG-98-311) (courtesy and copyright by COG).

for millions of years, because it is eventually quite close to the cooling of a molten volcano lava in oceans, seas, etc.

In addition, modern Generation-III$^+$ reactors are equipped with Passive-Core-Cooling System (PCCS), which at high heat flux will operate as boiling circulation loop (for details, see [41] or [29]).

More information on boiling, its characteristics, specifics, etc. can be found in the following publications: Boiling: Research and Advances [42]; Pioro et al. [43]; Naterer

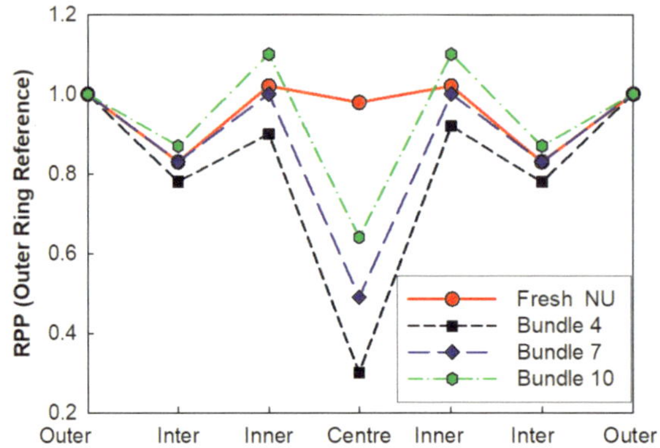

Figure 28.
Sample of radial power profiles (RPPs) for CANDU-reactor bundle used for critical heat flux (CHF) tests (based on report COG-98-311) (courtesy and copyright by COG).

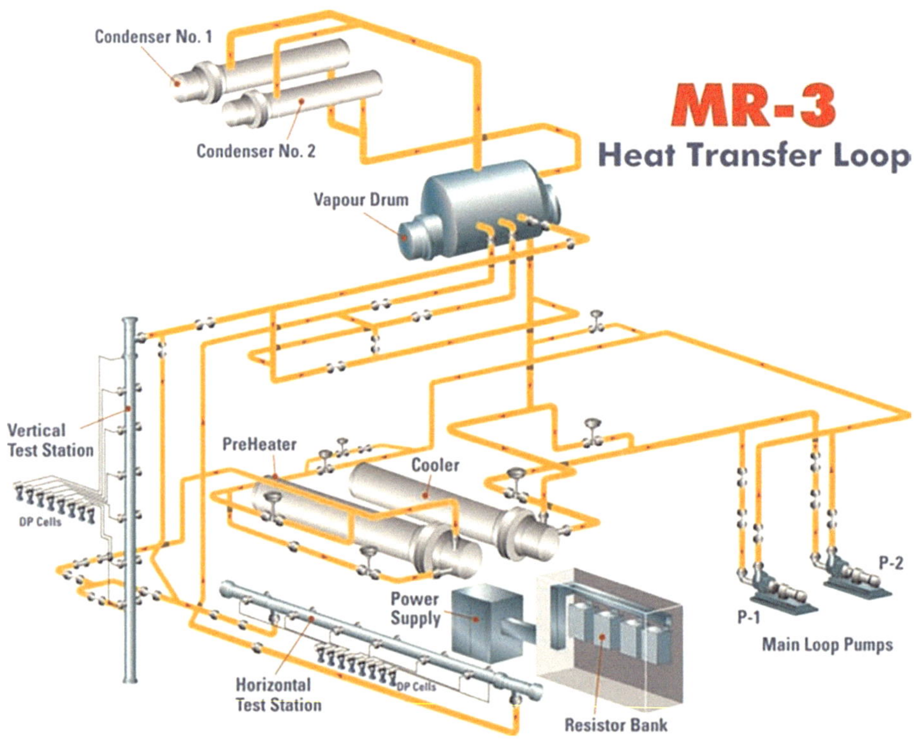

Figure 29.
Simplified layout of large (full-bundle-string) thermalhydraulics R-134a loop [39, 40].

[44]; Pioro et al. [45–47]; Handbook of Phase Change: Boiling and Condensation [48]; Groeneveld et al. [49]; Convective Flow Boiling [50]; Collier and Thome [51]; Lahey and Moody [52]; Whalley [53]; Hanne and Grigull [54]; Davis and Anderson [55]; Thorn et al. [56]; and Bergles and Rohsenow [57].

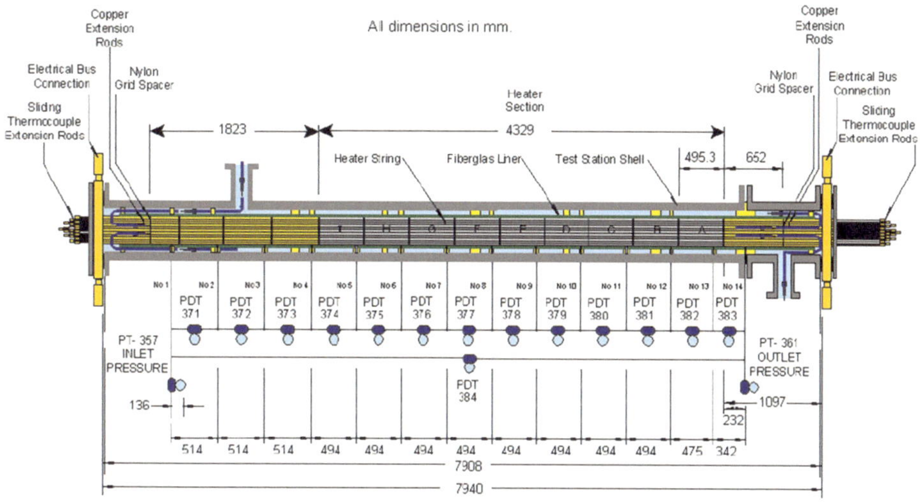

Figure 30.
General layout of horizontal test station of MR-3 R-134a thermalhydraulics loop [39, 40]: PDT—pressure differential transducer and PT—pressure transducer.

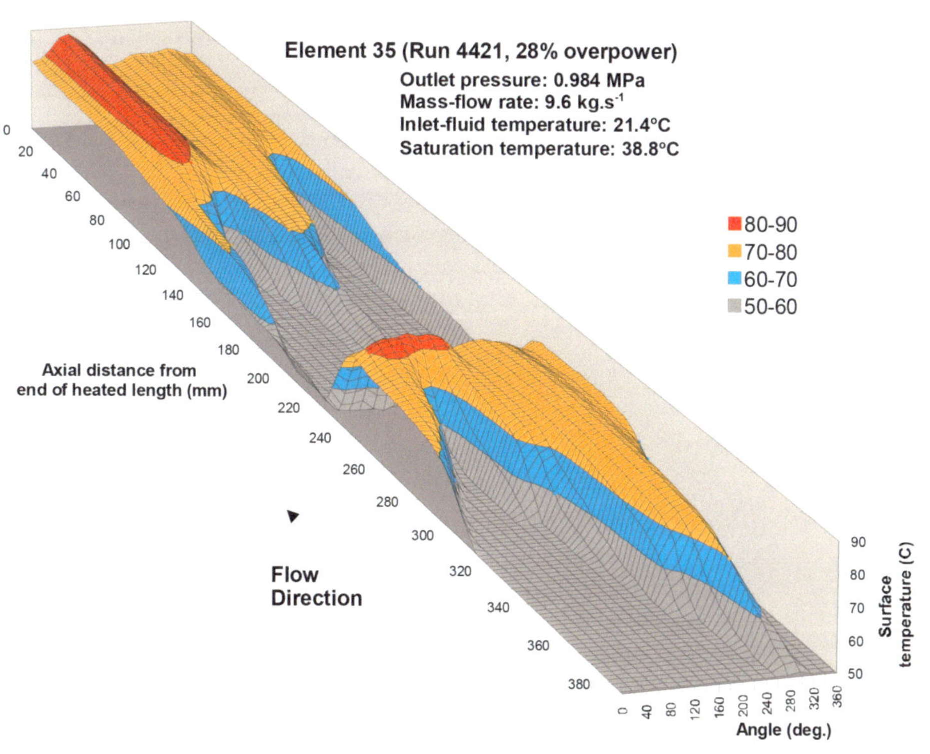

Figure 31.
Surface-temperature map for element 35 at 28% overpower (actual power to critical power): R-134a, 37-element bundle (for details, see [40]).

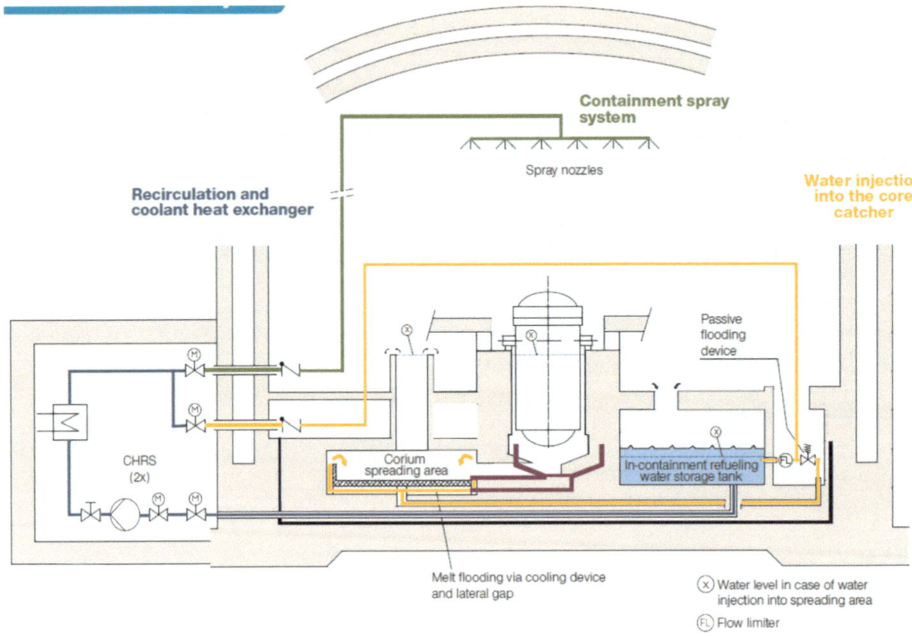

Figure 32.
Containment heat-removal system (CHRS) (courtesy and copyright by AREVA (EDF)). Two fully redundant trains with specific diversified heat sink.

5. Conclusion

In spite of more than 100 years of active research and even more years of applications, boiling phenomena/heat transfer are still not fully investigated and understood. There are some attempts to develop boiling-phenomena theories, but, unfortunately, they are not so practical yet. Therefore, more or less all practical calculations of various boiling characteristics/parameters rely heavily on empirical correlations, which were obtained experimentally. Due to this sophisticated studies are performed into boiling phenomena in the world.

Nomenclature

A	flow area, m^2
Al	Aluminum
c	specific heat, J/kg K
c_p	specific heat at constant pressure, J/kg K
D	inside diameter, m
D_b	vapor-bubble departure diameter, m
D_{hy}	hydraulic-equivalent diameter, m
f	frequency of vapor-bubbles departure, 1/s
G	mass flux, kg/m^{2}s; $\left(\frac{m}{A_{fl}}\right)$
g	gravitational acceleration, m/s^2
H, h	specific enthalpy, J/kg

k	thermal conductivity, W/m K
L	heated length, m
m	mass-flow rate, kg/s; ($\rho \cdot V$)
P, p	pressure, Pa
R	surface roughness parameters (for details, see Appendix A)
Q	heat-transfer rate, W
q	heat flux, W/m^2; $\left(\frac{Q}{A_h}\right)$
T, t	temperature, °C
$\overline{u}_{\mathrm{b}}$	mean velocity of vapor-bubble growth, ($D_{\mathrm{b}} \cdot f_{\mathrm{b}}$)
V	volume-flow rate, m^3/kg or volume, m^3
x	vapor quality

Greek Letters

Δ	difference
μ	dynamic viscosity, Pa·s
ρ	density, kg/m^3
σ	coefficient of surface tension (N/m)

Subscripts or superscripts

b	boiling
cr	critical
el	electrical
f	saturated fluid
fg	fluid-gas
fl	fluid
g	gas (saturated vapor)
h	heated
hor	horizontal
hy	hydraulic equivalent
in	inlet
lim	limited
max	maximum
min	minimum
sat	saturation
th	thermal
ver	vertical
w	wall

Abbreviations and acronyms widely used in the text and list of references

ABWR	Advanced Boiling Water Reactor
AECL	Atomic Energy of Canada Limited
AHFP	Axial Heat Flux Profile
Al	Aluminum
ASHRAE	American Society of Heating, Refrigerating and Air-Conditioning Engineers
Bi	Bismuth
CE	Common Era (the same as AD (Anno Domini); represents time from year 1 and onward)

CO_2	Carbon Dioxide
BWR	Boiling Water Reactor
CANDU	CANada Deuterium Uranium (reactor)
CHF	Critical Heat Flux
COG	CANDU Owners Group (Ontario, Canada)
Cu	Copper
EDF	Électricité de France S.A.
HE	Helium
hor.	horizontal
HTC	Heat Transfer Coefficient
ID	Inside Diameter
MHI	Mitsubishi Heavy Industries
NIST	National Institute of Standards and Technology (USA)
NPP	Nuclear Power Plant
NRC	Nuclear Regulatory Commission
NU	Natural Uranium
OD	Outside Diameter
ONB	Onset of Nucleate Boiling
OSV	Onset of Significant Void
PB	lead
PDO	Post-DryOut
PHD	Polyethylene High Density
PHWR	Pressurized Heavy-Water Reactor
Pt	Platinum
PWR	Pressurized Water Reactor
R	Refrigerant
RBMK	Reactor of Large Capacity Channel-type (in Russian abbreviations) cooled with boiling water)
REFPROP	REFerence PROPerties
SS, St.St.	Stainless Steel
USA	United States of America
vert.	vertical
W	Water

A. Appendix

Explanations to **Table A**.

Fluids are given in such order: water, alcohols, fluorocarbons (refrigerants by increasing in their number), hydrocarbons, and others.

No	Fluid-Surface	C_{sf}	n	T_{sat} °C	ΔT_b °C	q kW/m²	h_b kW/m²K
1	Water/Copper (plate oxidized, R_a = 1.37, R_q = 1.73)	0.015	0.81	23–82	4.2–14	1.8–72	0.26–6.1
2	Water/Copper (thin disk emery polished and paraffin treated)	0.0147	1	100	9.4–18.3	41–950	3.7–57
3	Water (internal forced convection)/ Copper (vertical tube)	0.013	1.7	100	4–15	18–190	4.5–12.7

No	Fluid-Surface	C_{sf}	n	T_{sat} °C	ΔT_b °C	q kW/m^2	h_b kW/m^{2}K
4	Water/Copper (disk emery polished) (thin copper disk (circular plate) (D = 25.4 mm, thickness 0.5 mm)	0.0128	1	100	5.5–15.5	28–530	3.4–34
5	Water/Copper (disk scored)	0.0068	1	100	1.7–8.3	1.6–190	8.5–19
6	Water/Aluminum (disk polished, R_q = 0.33)	0.011	1	100	5–8.3	28–126	5.6–15.1
7	Water/Aluminum (plate oxidized, R_a = 3.61, R_q = 4.52)	0.011	1.26	3.5–103	5–22	1.3–90	0.1–10
8	Water/Brass (plate, R_a = 0.47, R_q = 0.66)	0.015	0.81	14–103	4–15	1.2–144	0.2–11
9	Water/Brass (tube external)	0.009 0.006	1.1 1.7	28–112	2.3–14	8–43	1–11
10	Water/Chromium (polished thin layer on copper disk)	0.019	0.45	100–205	8–18	140–655	14.6–51
11	Water/Platinum (wire)	0.013	1	100–355	1–28	3–3200	1.8–480
12	Water/St. St. (plate, R_a = 0.75, R_q = 1.2)	0.015	0.69	30–103	4–13	5–40	1–6.2
13	Water/St. St. (disk polished, R_q = 0.13)	0.01	1	100	4.4–7.2	28–126	6.4–17.5
14	Water/St. St. (disk grounded, R_q = 0.53)	0.008	1	100	3.3–6.7	22–158	6.7–23.6
15	Water/St. St. (disk grounded, R_q = 3.6)	0.007	1	100	3.3–5.6	37–160	11.5–29
16	Water/St. St. (disk grounded and polished)	0.008	1	100	3–8.3	17–90	6–34
17	Water/St. St. (thin strip pitted with Teflon coating in pits	0.0058	1	100	0.5–8.3	1.6–265	3.1–31.8
18	Ethanol/Copper (plate oxidized, R_a = 1.37, R_q = 1.73)	0.00079	2.3	30–82	13–20.2	2.1–26	0.1–2
19	Ethanol/Aluminum(plate oxidized, R_a = 3.61, R_q = 4.52)	0.008	1.18	16–78	11–21	2.7–32	0.16–2.3
20	Ethanol/Brass (plate, R_a = 0.47, R_q = 0.66)	0.011	0.92	40–78	9–14	4.3–56	0.7–4.7
21	Ethanol/Chromium (polished thin layer on copper disk	0.0045 0.0027	1.47 1.7	78–210	3–34	15–800	3.2–72
22	Ethanol/St. St. (plate, R_a = 0.75, R_q = 1.2)	0.00053	2.28	49–90	7–14	10–45	0.7–6
23	Methanol/Brass (tube external boiling)	0.0026	1.7	13–72	5–19	7.8–43	0.7–6
24	iso-Propanol (internal flow boiling)/ Copper (vertical tube)	0.0022	1.7	82.5	5–20	5.7–95	1.1–4.8
25	n-Butanol (internal flow boiling)/ Copper (vertical tube)	0.03	1.7	117	7–17	9.5–95	1.4–5.6
26	R-11/Copper (plate oxidized, R_a = 1.37, R_q = 1.73)	0.0009	3.47	23–49	6.4–11.2	1.4–12	0.22–1.2
27	R-11/Copper (tube external) R_a = 0.17 R_a = 0.45 R_a = 2.30	0.0019 0.0014 0.00089	2.8	—	—	—	—

No	Fluid-Surface	C_{sf}	n	T_{sat} °C	ΔT_b °C	q kW/m^2	h_b kW/m^{2}K
28	R-11/St. St. (tube external) R_a = 0.02 R_a = 0.16	0.0031 0.0023	2.8	—	—	—	—
29	R-12/Copper (tube external)	0.016 [Pioro]	1.7	2–15	6.6–10	1.5–5	0.2–0.5
30	R-12/Copper (tube external) R_a = 0.06 R_a = 0.51	0.0031 0.00018	5	—	—	—	—
31	R-113/Copper (plate oxidized, R_a = 1.37, R_q = 1.73)	0.0022	2.25	32–80	9–16	3.2–21	0.24–1.9
32	R-113/Aluminum (plate oxidized, R_a = 3.61, R_q = 4.52); /Brass (plate, R_a = 0.47, R_q = 0.66); /St. St., (plate, R_a = 0.75, R_q = 1.2)	0.013	1.2	20–71	6–15	1.3–24	0.2–2
33	R-123/Copper (tube external) R_a = 0.16 R_a = 0.47 R_a = 3.30	0.0067 0.0055 0.0036	1.7	—	—	—	—
34	R-123/St. St. (tube external, R_a = 0.16)	0.0084	1.7	—	—	—	—
35	R-1234a/St. St. (tube external) R_a = 0.08	0.0025	2.5	—	—	—	—
36	R-134a/Copper (tube external) R_a = 0.07 R_a = 0.50 R_a = 2.50	0.0023 0.0016 0.00098	2.5	—	—	—	—
37	CCl_4 (internal flow boiling)/Copper (vertical tube)	0.013	1.7	76.7	10–28	5.5–63	0.55–2.3
38	CCl_4/ Copper (disk emery polished)	0.007	1.7	76.7	12–22	10.2–194	0.85–14
39	CCl_4/Copper (disk lapped)	0.0031	1.7	76.7	8–13	22–280	2.6–25
40	CCl_4/Brass (tube external)	0.0022	2.1	33–86	6.5–25	7.8–43	0.6–4
41	Propane/Chromium (polished thin layer on copper disk)	0.0069	2.17	34–83	3.3–22	27–400	8–28
42	n-Pentane/Copper (disk mirror finishing)	0.0171	1.7	36	16.7–44.4	22–252	1.3–5.7
43	n-Pentane/Copper (plate emery polished, R_q = 0.15)	0.0154	1.7	36	17–22	15–90	0.85–4
44	n-Pentane/Copper (disk emery rubbed)	0.0074	1.7	36	6.7–15.6	31.5–284	4.7–18.2
45	n-Pentane/Copper (disk lapped)	0.0049	1.7	36	5–8.3	31.5–284	6.3–34
46	n-Pentane/Zinc(disk polished)	0.0088	1.7	36	4–38	6.3–284	1.6–7.5
47	n-Pentane/Chromium (polished thin layer on copper disk)	0.015	1.7	48–147	4–31	29–440	2–35
48	n-Pentane/Nickel (disk mirror finishing)	0.0154	1.7	36	6.1–9.4	28–284	4.6–30.2
49	n-Pentane/Nickel (plate emery polished)	0.0127	1.7	36	11–22	13–112	0.85–5.7
50	n-Pentane/Nickel (disk lapped)	0.0043	1.7	36	15.5–39	31.5–284	2–7.3
51	n-Pentane/Inconel (disk mirror finishing)	0.018	1.7	36	19.4–43.3	31.5–221	1.6–5.1

No	Fluid-Surface	C_{sf}	n	T_{sat} °C	ΔT_b °C	q kW/m^2	h_b kW/m^{2}K
52	n-Pentane/Inconel (disk lapped)	0.0072	1.7	36	9.4–15.6	31.5–252	3.4–16.2
53	Benzene/Chromium (polished thin layer on copper disk)	0.01	1.7	80–214	4–45	25–600	2.5–41
54	n-Heptane/Chromium(polished thin layer on copper disk)	0.0014	1.37	95–185	10–32	40–440	2.8–25
55	Acetone/Copper (disk emery polished)	0.0096	1.7	56	14.4–22.2	16.4–252	1.1–11.4
56	35% K_2CO_3 (internal flow boiling)/ Copper (vertical tube)	0.0054	1.7	106	8–15	19–95	2.4–6.3
57	50% K_2CO_3 (internal flow boiling)/ Copper (vertical tube)	0.0028	1.7	72.4	10–17	25–95	2.5–5.6

Table A.
Average values of constants in the Rohsenow pool-boiling correlation and prediction intervals (for all references and other details of experiments and boiling surfaces/conditions, see [8]).

Materials of surfaces are located according to the value of thermal conductivity: from highest to lower ones.

Surfaces of the same material generally are located according to the decreasing value of C_{sf}.

All surfaces were located horizontally except where noted.

Generally two-phase thermosyphon-type chambers were used with boiling surface (plates, disks, or strips) located on the bottom or immersed in a pool (wires or tubes) and condensing part at the top.

In the present work plates (boiling surface 411 × 51 mm) from copper (no surface treatment, naturally oxidized, thickness 6.4 mm, R_a = 1.37, R_q = 1.73), aluminum (surface machined and oxidized, thickness 12.7 mm,), brass (no surface treatment, thickness 6.4 mm), and SS304 stainless steel (no surface treatment, thickness 9.62 mm) were used.

Author details

Igor L. Pioro
University of Ontario Institute of Technology, Oshawa, Ontario, Canada

*Address all correspondence to: igor.pioro@uoit.ca

References

[1] Bezrodny MK, Pioro IL, Kostyuk TO. Transfer processes in two-phase thermosyphon systems. In: Theory and Practice. 2nd ed. Kiev, Ukraine: Augmented and Revised, Fact Publ. House; 2005. 704 p (in Russian)

[2] Pioro LS, Pioro IL. Industrial Two-Phase Thermosyphons. New York, NY, USA: Begell House, Inc; 1997. 288 p

[3] NIST REFPROP. Lemmon EW, Bell IH, Huber ML, McLinden MO, editors. Reference Fluid Thermodynamic and Transport Properties, NIST Standard Reference Database 23, Ver. 10.0. National Institute of Standards and Technology, USA. 2018. Available from: https://www.nist.gov/srd/refprop [Accessed: December 10, 2023]

[4] Pioro I, Mahdi M, Popov R. Heat transfer media and their properties. In: Kulacki FA, editor. Chapter 33 in: Handbook of Thermal Science and Engineering. Cham, Switzerland: Springer; 2018. pp. 1353-1446

[5] Pioro IL, Kirillov PL, Razumovskiy VG, Pis'mennyy EN. Appendix A4. Current Status of Research on Thermalhydraulics in Forced Convection of Fluids at Supercritical Pressures in Bare Tubes, Annular- and Bundle-Flow Geometries. 2023. 40 p.

[6] Pioro IL, Mokry SJ, Dort-Goltz N, McKellar J, Xie H. Appendix A3. Thermophysical Properties of Supercritical Fluids. 2023. 84 p.

[7] Pioro IL, Duffey RB. Heat Transfer and Hydraulic Resistance at Supercritical Pressures in Power Engineering Applications. New York, NY, USA: ASME Press; 2007. 334 p

[8] Pioro IL. Experimental evaluation of constants for Rohsenow pool boiling correlation. International Journal of Heat and Mass Transfer. 1999;**42**(11): 2003-2013

[9] Bergman TL, Lavine AS, Incropera FP, DeWitt PD. Fundamentals of Heat and Mass Transfer. 8th ed. Hoboken, NJ, USA: J. Wiley & Sons; 2017. 992 p

[10] Pioro IL, Rohsenow W, Doerffer S. Nucleate pool-boiling heat transfer—II. Assessment of prediction methods. International Journal of Heat and Mass Transfer. 2004;**47**(23):5045-5057

[11] Köhler W, Hein D. Influence of the wetting state of a heated surface on heat transfer and pressure loss in an evaporator tube. In: International Agreement Report NUREG/IA–0003, 1986. 1986. pp. 63-79. Available from: https://www.nrc.gov/docs/ML0628/ML062840385.pdf [Accessed: December 10, 2023]

[12] Fahri A. Heat Pipe Science and Technology. 2nd ed. Columbia, MO, USA: Global Digital Press; 2016. 1013 p

[13] Reay D, McGlen RJ. Heat Pipes: Theory, Design and Applications. 6th ed. Oxford, UK: Elsevier; 2014. 251 p

[14] Petersen GP. An Introduction to Heat Pipes. New York, NY, USA: J. Wiley & Sons, Inc.; 1994. 356 p

[15] Pioro IL, Lee Y. Critical heat flux in two-phase thermosyphons with binary mixtures. In: Proceedings of the 3rd International Symposium on Multiphase Flow and Heat Transfer; 19–21 September 1994; Xi'an, China. Vol. 1. 1994. pp. 46–55

[16] Tikhonovskiy SA, Pioro IL. Maximum heat flux to antifreeze boiling in a two-phase thermosiphon. Heat Transfer Research. 1992;**24**(3):390-397

[17] Webb RL, Kim N-H, Webb RL. Principles of Enhanced Heat Transfer. 2nd ed. New York, NY, USA: Taylor & Francis; 2005. 795 p

[18] Kraus AD, Aziz A, Welty J. Extended Surface Heat Transfer. New York, NY, USA: J. Wiley & Sons, Inc.; 2001. 1105 p

[19] Thome JR. Enhanced Boiling Heat Transfer. New York, NY, USA: Hemisphere Publ. Corp; 1990. 356 p

[20] Bystrov PI, Kagan DN, Krechetova GA, Shpilrain EE. Liquid-Metal Coolants for Heat Pipes and Power Plants. New York, NY, USA: Hemisphere Publ. Corp; 1990. 272 p

[21] Dwyer OE. Boiling Liquid-Metal Heat Transfer. La Grange Part, IL, USA: American Nuclear Society (ANS); 1976. 446 p

[22] Weisend JG II, editor. Handbook of Cryogenic Engineering. Philadelphia, PA, USA: Taylor & Francis; 1998. 504 p

[23] Pioro IL, Rohsenow W, Doerffer S. Nucleate pool-boiling heat transfer—I. Review of parametric effects of boiling surface. International Journal of Heat and Mass Transfer. 2004;**47**(23): 5033-5044

[24] Tolubinskiy VI. Heat Transfer at Boiling. Kiev, Ukraine: Naukova Dumka; 1980. 315 p (in Russian)

[25] Kichigin AM, Pioro IL. Major approaches to solving the boiling crisis problem under free flow conditions (a review). Heat Transfer Research. 1995; **26**(1 & 2):82-90

[26] Kichigin AM, Pioro IL. Analysis of methods for detecting Dryout and Burnout fluxes. Heat Transfer Research. 1992;**24**(7):957-964

[27] Kichigin AM, Pioro IL. About crisis mechanism of heat transfer under boiling in conditions of free liquid motion. Promyshlennaya Teplotekhnika. 1991;**13**(3):19-33 (in Russian)

[28] Hewitt GF, Shires GL, Bott TR. Process Heat Transfer. Boca Raton, FL, USA: CRC Press; 1994. 1042 p

[29] Pioro IL, editor. Handbook of Generation IV Nuclear Reactors. 2nd ed. Kidlington, UK: Elsevier, Woodhead Publishing (WP); 1079 p (hard copy) and 197 p. (Appendices 3–9 on website). Available from: https://shop.elsevier.com/books/handbook-of-generation-iv-nuclear-reactors/pioro/978-0-12-820588-4;: Elsevier–Woodhead Publishing (WP); 2023 https://www.gen-4.org/gif/jcms/c_208948/see-link-for-further-information [Accessed: December 10, 2023]

[30] Kirillov PL, Zhukov AV, Loginov NI, Makhin VM, Pioro IL, Yur'ev YS. Nuclear reactors, heat exchangers, steam generators. In: Kirillov PL, editor. Handbook on Thermalhydraulics Calculations in Nuclear Engineering. Vol. 2. Moscow, Russia: IzdAT Publishing House; 2013. 688 p (In Russian)

[31] ASHRAE Handbook: Fundamentals. 2021. Available from: https://www.ashrae.org/technical-resources/ashrae-handbook/description-2021-ashrae-handbook-fundamentals [Accessed: December 10, 2023]

[32] McQuiston FC, Parker JD. Heating Ventilating, and Conditioning: Analysis and Design. New York, NY, USA: J. Wiley & Sons, Inc.; 1994. 742 p

[33] Woodruff EB, Lammers HB, Lammers TF. Steam Plant Operation. 10th ed. New York, NY, USA: McGraw Hill; 2017. 750 p

[34] Rayaprolu K. Boilers for Power and Process. Boca Raton, FL, USA: CRC Press; 2009. 745 p

[35] Hewitt GF, Collier JG. Introduction to Nuclear Power. 2nd ed. New York, NY, USA: Taylor & Francis; 2000. 304 p

[36] Pioro IL, Groeneveld DC, Cheng SC, Doerffer S, Vasić A, Antoshko YV. Comparison of CHF measurements in R-134a cooled tubes and the water CHF look-up table. International Journal of Heat and Mass Transfer. 2001;**44**(1): 73-88

[37] Pioro IL, Groeneveld DC, Leung LKH, Doerffer SS, Cheng SC, Antoshko YV, et al. Comparison of CHF measurements in horizontal and vertical tubes cooled with R-134a. International Journal of Heat and Mass Transfer. 2002; **45**(22):4435-4450

[38] Pioro IL, Groeneveld DC, Doerffer SS, Guo Y, Cheng SC, Vasic A. Effects of flow obstacles on the critical heat flux in a vertical tube cooled with upward flow of R-134a. International Journal of Heat and Mass Transfer. 2002; **45**(22):4417-4433

[39] Guo Y, Bullock DE, Pioro IL, Martin J. Measurements of sheath temperature profiles in Bruce LVRF bundles under post-dryout heat transfer conditions in Freon. In: Proceedings of the 14th International Conference on Nuclear Engineering (ICONE-14); 17–20 July; Miami, Florida, USA. 2006. Paper #89621, 9 p

[40] Leung LKH, Pioro IL, Bullock DE. Post-dryout surface-temperature distributions in a vertical Freon-cooled 37-element bundle. In: Proceedings of the 10th International Topical Meeting on Nuclear Reactor Thermal Hydraulics (NURETH-10); 5–9 October; Seoul, Korea. 2003. Paper #C00201, 13 p

[41] Friedman BN. The AP1000 plant and the China project commissioning success. Journal of Nuclear Engineering and Radiation Science. 2019;**5**(1): 3 p

[42] Koizumi Y, Shoji M, Monde M, Takata Y, Nagai N, editors. Boiling: Research and Advances. Oxford, UK and Japan: Elsevier ans JSME; 2017. 801 p

[43] Pioro I, Duffey R, Dumouchel T. Hydraulic resistance of fluids flowing in channels at supercritical pressures (survey). Nuclear Engineering and Design. 2004;**231**(2):187-197

[44] Naterer GF. Heat Transfer in Single and Multiphase Systems. Boca Raton, FL, USA: CRC Press; 2003. 618 p

[45] Pioro IL, Cheng SC, Vasić A, Felisari R. Some problems for bundle CHF prediction based on CHF measurements in simple flow geometries. Nuclear Engineering and Design. 2000;**201**(2–3):189-207

[46] Pioro IL, Cheng SC, Groeneveld DC, et al. Experimental study of the effect of non-circular flow geometry on the critical heat flux. Nuclear Engineering and Design. 1999;**187**:339-362

[47] Pioro IL, Cheng SC, Vasić A, Salah I. Experimental evaluation of the limiting critical quality values in circular and non-circular geometries. Nuclear Engineering and Design. 1999;**190**: 317-339

[48] Kandlikar SG, Dhir VJ, editors. Handbook of Phase Change: Boiling and Condensation. Philadelphia, PA, USA: Taylor & Francis; 1999. 738 p

[49] Groeneveld DC, Leung LKH, Kirillov PL, et al. The 1995 look-up table for critical heat flux in tubes. Nuclear Engineering and Design. 1996, 1996;**163**: 1-23

[50] Chen JC, editor. Convective Flow Boiling. Washington, DC, USA: Taylor & Francis; 1995. 391 p

[51] Collier JG, Thome JR. Convective Boiling and Condensation. 3rd ed. Oxford, UK: Clarendon Press; 1994. 640 p

[52] Lahey RT, Moody FJ. The Thermal-Hydraulics of a Boiling Water Nuclear Reactors. 2nd ed. La Grange Part, IL, USA: American Nuclear Society (ANS); 1993. 631 p

[53] Whalley PB. Boiling Condensation and Gas-Liquid Flow. Oxford, UK: Clarendon Press; 1987. 291 p

[54] Hanne E, Grigull U. Heat Transfer in Boiling. New York, NY, USA: Hemisphere Publ. Corp; 1977. 486 p

[55] Davis EJ, Anderson GH. The incipience of nucleate boiling in forced convection flow. AICHE Journal. 1966; **12**(4):774-780

[56] Thorn JRS, Walker WM, Fallon TA, Reising GFS. Boiling in subcooled water during flow up heated tubes or annuli. Proceedings of the Institution of Mechanical Engineers. 1965;**180**:226-246

[57] Bergles AE, Rohsenow WM. The determination of forced convection surface-boiling heat transfer. ASME Journal of Heat Transfer. 1964;**86**: 365-372

Chapter 2

Heat Transfer and Hydraulic Resistance in Nuclear Fuel Rods

Armando Nava Dominguez

Abstract

This chapter briefly describes the fundamental concepts of heat transfer and hydraulic resistance in water-cooled nuclear reactors, more specifically the nuclear fuel assemblies. There are two key areas in nuclear thermal-hydraulics, namely heat transfer and hydraulic resistance with and without phase change. Boiling and condensation play a critical role in water-cooled nuclear reactors as these are needed for the design, operation and safety analysis of nuclear reactors. The common models used in the nuclear industry are described in this chapter.

Keywords: boiling, two-phase flow, nuclear thermal-hydraulics, hydraulic resistance, flow patterns

1. Introduction

The word thermal-hydraulics is composed of two Greek terms which indicate heat and water respectively. Therefore, thermal-hydraulics studies the behaviour of fluid (s) subjected to heat. In particular, nuclear thermal-hydraulics is more specific to the analysis of nuclear systems, such as nuclear power plants (NPPs). It is used mainly for the design of thermal systems, dimensioning of thermal or hydraulic components, system performance and safety analysis. Key phenomena associated with this discipline are boiling heat transfer, hydraulic resistance under single-phase and multi-phase conditions.

The field of nuclear thermal-hydraulics, specifically for water-cooled reactors, improved significantly during the peak of the nuclear industry. This resulted in the production and publication of a large body of knowledge, such as books, articles, conference proceedings. It would be impossible, and impractical, to replicate all this information in a single chapter. For that reason, this chapter entitled "Heat transfer and hydraulic resistance in nuclear fuel rods" presents only the fundamentals of nuclear thermal-hydraulics, with a focus on CANDU (CANada Deuterium Uranium) type reactors. For that reason, the model developments and equations are omitted. The reader is instead led to make use of the references cited in this chapter for further details.

2. Reactor power

Important input needed for thermal-hydraulics analysis is the core power distribution, that is, how the power is distributed within a reactor's core. This analysis is usually determined by carrying out reactor physics calculations. In CANDU reactors, the fuel management differs from pressure vessel water-cooled reactors, because it uses natural uranium, which requires a different fuel management process. On-power refuelling is used to replace fuel as needed, to keep the reactor operating as designed. Furthermore, CANDU reactors use independent fuel channels to contain their nuclear fuel, and consequently, each fuel channel produces different powers.

To start the reactor power analysis, the link between the neutron flux with power is represented by the following equation:

$$q'''(r) = E_d \int_0^{\infty} \Sigma_f(E)\Theta(r,E)dE \tag{1}$$

Where E_d is the energy deposited locally in the fuel per fission, Σ_f is the macroscopic fission cross-section of the fuel, and Θ is the neutron flux. This equation essentially means that power generated per unit of volume is the neutron reaction rate times the energy deposited per fission. For a cylindrical homogeneous reactor without a reflector, the neutron flux can be approximated [1].

$$\Theta(r,z) = A\, J_o \left[\frac{2.405\, r}{R_e}\right] cos \left[\frac{\pi z}{H_e}\right] \tag{2}$$

Where:

$$A = \frac{3.63P}{VE_R\Sigma_f} \tag{3}$$

R_e and H_e are the extrapolated lengths of the radius and height of the cylinder, and J_o is the Bessel function of order zero, E_R is the recoverable energy in joules per fission and V the volume of the cylinder.

After some mathematical manipulations, the rate of heat production per unit of volume of a fuel rod is [1, 2]:

$$q'''(r,z) = q'''_{center} J_o \left[\frac{2.405\, r}{R_e}\right] cos \left[\frac{\pi z}{H_e}\right] \tag{4}$$

The solution of this equation is depicted in **Figure 1**. This figure shows the axial (z) and radial (r) power distribution.

In reality, the power distribution is an implicit calculation between neutron physics and thermal-hydraulics. In order to compute the neutron flux and power, it is necessary to know the coolant and fuel temperature distributions, which are unknown until thermal-hydraulics analyses are performed. However, the latter analyses require power and power distribution, and this results in an implicit calculation. Usually, a coupled and iterative process is used to get a solution. The reader is referred to reference [2] for additional information.

From Eq. (4), it can be stated that the power distributions are non-uniform, in fact, these have a cosine-like profile. The fact that the radial (core) power distribution

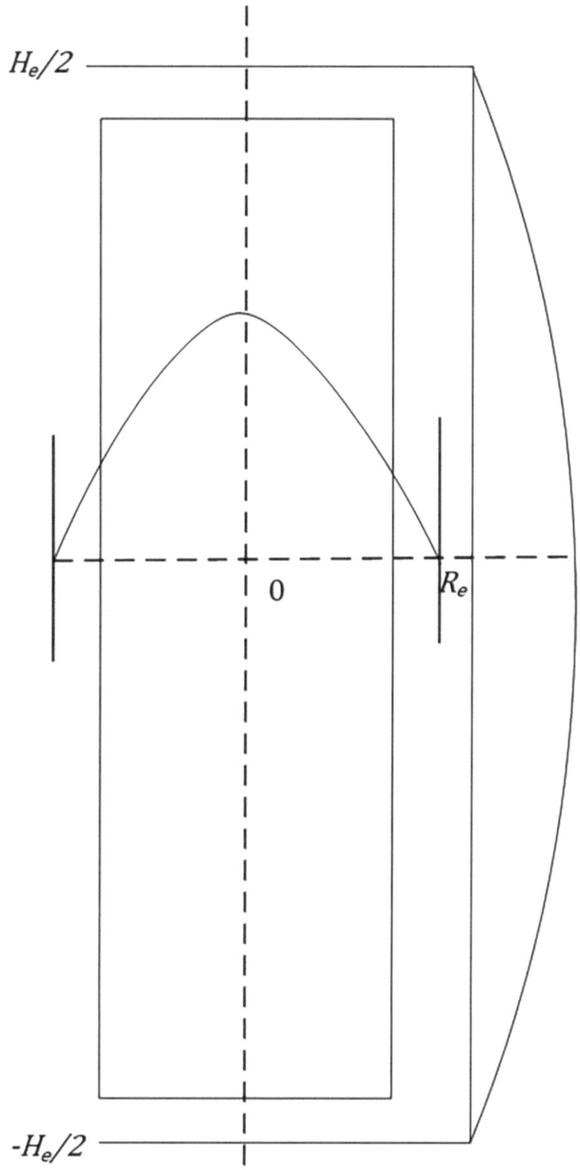

Figure 1.
Representation of an axial and radial power distribution.

is non-uniform has important effects on the dimensioning of equipment, systems and/or components of a nuclear core. For example, in CANDU reactors some fuel channels produce more power than others. The ones near the centre usually produce more than those that are in the periphery. Furthermore, as the inlet and outlet coolant temperatures are fixed for all the fuel channels, it is important to ensure each fuel channel has received the right amount of coolant to remove the heat generated by the nuclear fuel. Indeed, fuel channels with lower power require less mass flow than fuel channels with higher power. In CANDU reactors, the channel mass flow is not a controlled parameter. Orifices are used in the fuel channel to provide the right amount of coolant in the fuel channels. However, under transient conditions or accidents, the channel flow varies with changes in the reactor and channel powers, and with the thermal-hydraulics conditions at the inlet and outlet headers.

As CANDU reactors are refuelled online, the insertion of fresh bundles in a channel, which usually replaces four bundles at a time-creates a local-power effect due to the inserted reactivity (see **Figure 2**). This is called ripple effect and can impact several neighbouring fuel channels. In addition, the continuously changing burn up distribution can result in power oscillations due to Xenon or regional overpower[1]. These overpower limits are separate from and above the normal operating limits on channel and bundle powers.

Another important characteristic of CANDU fuel channels is the End-Flux-Peaking (EFP). This is a phenomenon that affects the flux profile of a fuel bundle. It occurs in the end region separating two individual bundles. The geometry in the end regions consists of D_2O coolant, a Zircaloy endplate and end caps, and uranium dioxide fuel pellets. Due to the fact that the heavy-water-coolant and Zircaloy structural material have a much lower absorption cross-section for thermal neutrons than uranium dioxide, thermal neutrons tend not to be absorbed as much as they would be in the fuel. The lack of absorption of thermal neutrons in these regions leads to peaks in neutron flux around the end regions. The occurrence of EFP leads to higher fission rates, leading to more heat being produced and therefore overall higher temperatures within the pellets that are adjacent to these end regions. Higher temperatures lead to an increased risk of sheath strain, corrosion, and fuel centreline melting, all of which have a significant impact on the integrity of the sheath and fuel pellets during an overpower of loss-of-coolant-accident conditions.

2.1 Heat characteristics of a CANDU 6 nuclear reactor

In a nuclear reactor, the heat produced by the nuclear reaction is primarily removed through conduction and convection. For example, the heat produced in the nuclear fuel is transferred by conduction to the surface of the rod. Convection

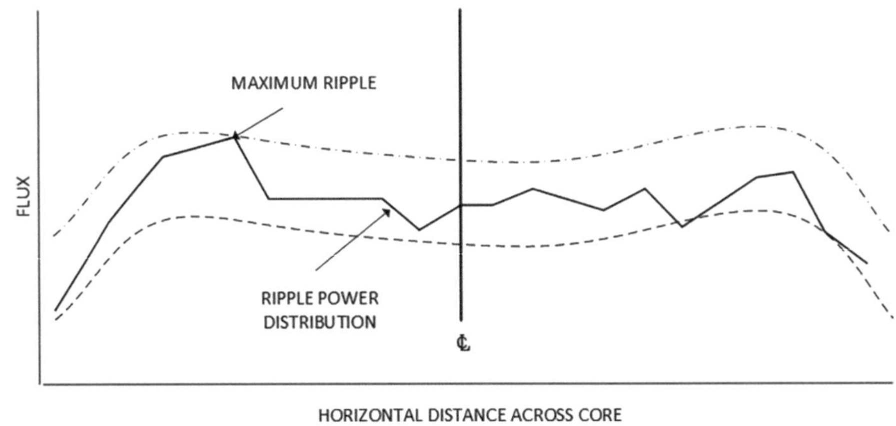

Figure 2.
Schematic of a CANDU power distribution.

[1] Overpower is defined as a fuel bundle or channel power in excess of specified safety-related limits.

involves the transfer of heat by the movement of a fluid. Thus, the heat conducted to the surface of a fuel rod is transferred to the coolant by convection. In addition, under some specific accidents, where the fuel reaches high temperatures, radiation heat transfer plays an important role.

2.1.1 Heat conduction equation

Fortunately, as the reactor core, fuel channels and fuel elements have a long cylindrical shape, the heat conduction equation can be written in cylindrical coordinates, which can then be used as a base for the heat conduction analysis. The general heat conduction equation is:

$$\rho C_p(r, T)\frac{\partial T}{\partial t} = \nabla \bullet k(T)\nabla T + q''' \tag{5}$$

Where C_p is the specific heat at constant pressure (which for incompressible materials is equal to the specific heat and constant volume C_v), ρ is the density of the fuel, k is the thermal conductivity, T is the temperature and q''' is the volumetric heat.

Note that the conductivity of the fuel is temperature dependent, as shown in Eqs. (6) and (7), which are formulations for the thermal conductivity of Uranium [3]:

$$k = \frac{100}{6.548 + 23.533T} + \frac{6400}{T^{5/2}}\exp\left(\frac{-16.35}{T}\right) \tag{6}$$

Where T is defined as $T/1000$, and T is in Kelvin. The conductivity k is the thermal conductivity for 95% dense UO_2 in W/m K.

Eq. (6) can be approximated using polynomials [3]:

$$\begin{aligned} k = &+12.57829 - 2.31100 \times 10^{-2}T + 2.36675 \times 10^{-5}T^2 - 1.30812 \times 10^{-8}T^3 \\ &+ 3.63730 \times 10^{-12}T^4 - 3.90508 \times 10^{-16}T^5 \end{aligned} \tag{7}$$

To better understand the analysis of the heat conduction equation, the following example is that of a classical heat transfer problem used in nuclear reactor analyses, which is the analysis of a single fuel element (see **Figure 3** for details). The objective is to solve the heat conduction equation, in order to determine the temperature profile within the fuel element. Both the maximum centreline temperature and cladding temperature are needed to ensure the integrity of the fuel and cladding is not

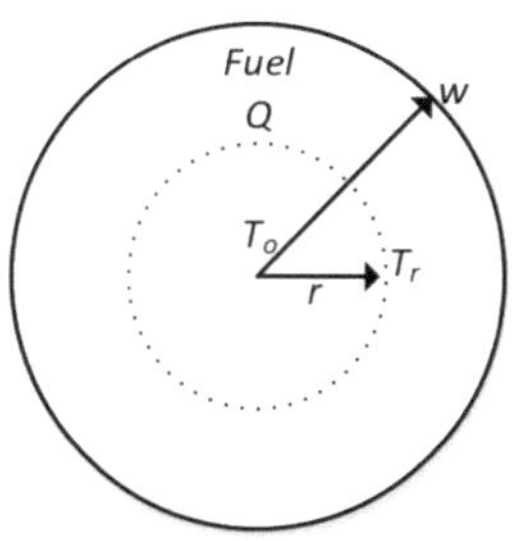

Figure 3.
Cross-section geometry of a fuel element.

compromised. The average temperature of the fuel is also needed for nuclear physics calculations. For example, assuming a predominant radial heat transfer under steady-state conditions, which is common in a nuclear fuel rod because the cylinder length is several times larger than the diameter, Eq. (5) is recast as:

$$\frac{1}{r}\frac{\partial}{\partial r}\left(r\,k\frac{\partial T}{\partial r}\right)+q'''=0 \tag{8}$$

Two boundary conditions are needed to solve this equation:

1. A boundary condition specifying that the heat flux at the centre of the fuel element is zero, as it is axisymmetric with respect to the axial direction,

2. The description of the heat transfer between the bulk temperature of the fluid (T_b) and the outer cladding (wall) temperature of the element (T_w) using Newton's law of cooling.

$$\left.\frac{\partial T}{\partial r}\right]_{r_0=0}=0\ for\ r=0 \tag{9}$$

$$q''=-k\left.\frac{\partial T}{\partial r}\right]_{r=w}=\tilde{h}(T_w-T_b)\quad for\quad r=w \tag{10}$$

Fortunately, this equation is not difficult to solve. For simplified cases, it is possible to find an analytical solution. As the problem is more elaborate, the equation is frequently solved by using numerical methods. In reality, the problem is much more challenging, as the materials—such as the nuclear fuel—undergo several nuclear and chemical reactions that change their properties over time. In addition, the mechanical properties are also affected by the conditions to which the materials are exposed during their lifetime. Changes in geometry are also possible, such as element bowing or ballooning. In other cases, oxide deposits (crud) in the cladding need to be considered.

The nuclear industry uses specific codes to assess in detail the heat transfer from the fuel element to the coolant. These codes must take into account the change of properties of the materials over time as the fuel is consumed and fission products are produced inside the fuel. Chemical reactions and mechanical stresses that can change the geometry of the system should also be considered for a complete analysis of the fuel element.

2.2 Heat transfer to coolants

Thus far, the value of the heat transfer coefficient $\left(\tilde{h}\right)$ has not been discussed in detail. Calculating and/or accurately estimating this coefficient is one of the most important and more difficult parts in thermal-hydraulics analyses. The reason for that is that this coefficient depends on multiple variables, such as geometry, flow conditions, flow regime and coolant properties.

Note that the second boundary condition needs the value of $\tilde{h}$. This boundary condition links the conduction with the convection heat transfer mechanisms. In other words, the heat from a solid is removed by a fluid (liquid or gas). This is described according to Newton's law of cooling [1]:

$$q'' = \tilde{h}(T_b - T_w) \tag{11}$$

Where $q^{"}$ is the heat flux, T_w is the temperature of the surface of the solid (in the case of reactor fuel, this is the outer temperature of the cladding or wall) and T_b is the temperature of the fluid.

The numerical value of the heat transfer coefficient, $\tilde{h}$ is a function of several variables such as physical properties of the fluid, mass flow rate, geometry and orientation of the system.

The solution of Newton's law of cooling for single-phase conditions has been commonly solved using the method of similarities, or dimensional analysis [4]. Three dimensionless numbers are used to correlate the heat transfer coefficient, namely the Reynolds (**Re**), the Nusselt (**Nu**) and Prandtl (**Pr**) numbers

$$\mathbf{Re} \equiv \frac{GD_e}{\mu} \tag{12}$$

$$\mathbf{Pr} \equiv \frac{C_p\mu}{k} \tag{13}$$

$$\mathbf{Nu} \equiv \frac{\tilde{h}D}{k} \tag{14}$$

The solution takes the form of:

$$\mathbf{Nu} = a\,\mathbf{Re}^{b}\mathbf{Pr}^{c} \tag{15}$$

Where a, b and c are coefficients, which are usually obtained experimentally.

For example, one of the most common correlations is the Dittus-Boelter correlation [5]:

$$\frac{D\tilde{h}}{k} = \mathbf{Nu} = 0.023\,\mathbf{Re}^{0.8}\mathbf{Pr}^{0.4} \tag{16}$$

Another common correlation is the Sieder and Tate correlation [6]

$$\frac{D\tilde{h}}{k} = \mathbf{Nu} = 0.023\,\mathbf{Re}^{0.8}\mathbf{Pr}^{\frac{1}{3}}\left(\frac{\mu_b}{\mu_w}\right)^{0.14} \tag{17}$$

Both correlations are applicable under single-phase conditions. However, the Sieder and Tate correlation [6] (Eq. (17)) uses an additional term to take into account the effect of the wall temperature, by also taking into account the viscosity evaluated at the wall temperature.

The boiling process is a clear example of a thermal-hydraulics analysis. It is evident that this process involves fluid dynamics and heat transfer. Fluid dynamics are needed to determine the momentum and enthalpy characteristics of the flow, such as velocity, pressure drop and void profiles. Heat transfer is also required to compute the temperature distributions of the system and heat transfer rates. The calculations are highly inter-related as one feeds back to the other. For example, the velocity of the flow affects the heat transfer rate, which in turn affects void generation (boiling) affecting the hydraulic resistance, which in turn can change the flow regime, and so forth, as these disciplines are all highly coupled. The next Section 3 briefly explains the forced flow boiling process.

3. Boiling heat transfer

As CANDU reactors are water-cooled, it is important to briefly describe the boiling process. This process is usually divided into two types: (1) pool boiling, in which there is no liquid mass flow entering or leaving the system, similar to that of an actual pool, and (2) the second deals with systems where the liquid mass flow enters and leaves the system. In a CANDU power plant, these two types of analysis are relevant. However, in this chapter, we focus solely on the analysis of a CANDU fuel channel. In a CANDU fuel channel, the coolant mass flow is forced as it is driven by the main heat transport pumps. As the liquid coolant moves through the fuel channel, the liquid absorbs the heat produced by the fuel bundles/elements from the nuclear reaction. As the coolant moves along the channel, it develops velocity and temperature distributions. These distributions dictate the process of boiling.

3.1 Boiling curve (forced flow)

Understanding the phenomenon of heat transfer under boiling conditions, or heat transfer with phase change, is critical in thermal-hydraulics analyses. A detailed review of the boiling process is provided in several nuclear engineering and convection heat transfer textbooks [7–18].

The boiling process can be explained by assuming a system where a heat flux from a heated fuel rod is transferred to a liquid coolant, such as water. The results are measured as a function of the temperature of the surface of the rod for a given system pressure and flow rate. The results of this experiment are shown in **Figure 4**.

The heat flux increases slowly as the rod temperature is increased at low values. In this temperature range, between points A and B, heat is transferred to the coolant by convention without phase change. The heat transfer coefficient is usually determined by empirical correlations, such as Eqs. (16) and (17).

As the surface temperature of the fuel is increased further, a point is eventually reached where bubbles or vapour form at various imperfections on the surface of the fuel rod. This occurs at about point B (in **Figure 4**). This type of boiling is called *nucleate boiling*. As the bubbles are formed, they are entrained from the rods into the bulk of the coolant as a result of the turbulent movement of the fluid. However, if the bulk temperature of the coolant is lower than its saturation temperature, the vapour condenses in the liquid disappearing from the coolant. Thus, there is no net production of steam under these circumstances. This boiling process is known as *subcooled nucleate boiling* or *local boiling*. If and when the bulk temperature of the coolant reaches its operation temperature, the bubbles remain within the coolant stream, there begins to be a net production of steam and the system is undergoing a *saturated nucleate boiling* or *bulk boiling*.

In any event, with the onset of nucleate boiling, the heat moves into the liquid. At every temperature in this region between B and C, heat transfer is more efficient than ordinary convection. There are two reasons for this. First, heat is removed from the rods both as heat of vapourization and sensible heat. Second, the motions of the bubbles lead to rapid mixing of the fluid. The rapid increase in the heat flux with

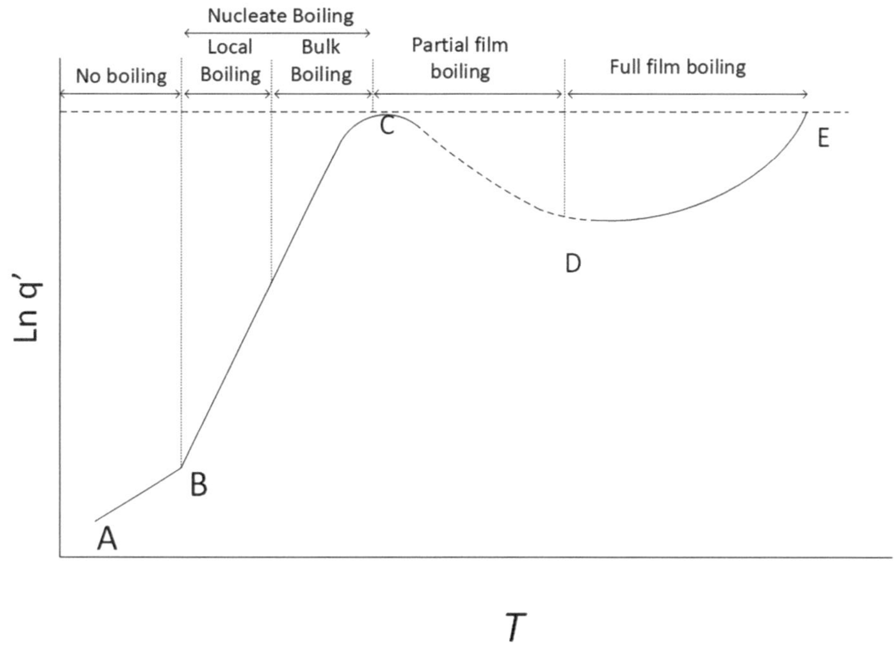

Figure 4.
Boiling curve.

temperature is explained by the fact that the density of the bubbles forming at and departing from the rod surface increases rapidly with surface temperature.

With increased vapourization in a coolant channel, the heated surface becomes intermittently exposed to patches of vapour. Since the heat transfer coefficient decreases when the surface is covered with vapour, the wall temperature rises correspondingly. Hence, the wall temperature may become unstable as the surface is alternately covered with vapour or liquid, but then rise after the wall liquid is completely vapourized. Such behaviour, characterised by a marked temperature rise of the heated surface during boiling, as a result of a change in the heat-transfer mechanism, is called *boiling crisis*. There are two types of boiling crisis, *departure of nucleate boiling* or *DNB* and *dryout*.

With the onset of the boiling crisis, the heat flux into the coolant begins to drop. This is due to the fact that over the regions of the rods covered by vapour film, the heat is forced to pass through the vapour into the coolant by conduction and radiation, both are relatively inefficient heat transfer mechanisms. The heat transfer continues to drop more or less erratically (dotted line) with increasing fuel temperatures as the total area of the film covering the fuel increases. In this region, the system is said to be experiencing partial film boiling.

Eventually, when the rod surface temperature is high enough, the vapour film covers the entire rod and the heat flux to the coolant falls to a minimum value (point D). Beyond this point, any increase in temperature leads to an increase in the heat flux simply because heat transfer through the film, although a poor and inefficient process, nevertheless increases with the temperature difference across the film. The system is said to be undergoing full film boiling.

The critical heat flux, at which burnout is expected to occur, is an important design consideration in water-cooled reactors. The knowledge of burnout conditions is important not only for the design of a fuel bundle and its maximum operating conditions at nominal power but also under upset conditions, such as might arise from loss of coolant flow conditions due to power excursions.

4. Fluid dynamics

As previously mentioned in Section 2, the energy (heat) generated in the nuclear reactor core is removed by the coolant mass flow. The thermal analysis of a nuclear power plant requires the knowledge of the pressure and velocity distribution of the coolant at different locations of the reactor, and the transfer of heat between fuel rods to the coolant, or between the primary and secondary sides.

There is a considerable amount of literature on fluid dynamics. For that reason, only the equations needed to explain a relevant topic in a nuclear power plant are hereafter provided. The solution or procedures to solve these equations are beyond the scope of this chapter.

4.1 Single-phase flow

Various forms of the single-phase fundamental equations exist. This chapter follows the format used by Delhaye et al. [15]. In this subsection, we present the fundamental conservation equations. These equations can be written in a general form as:

$$\int_{V(t)} \frac{\partial \varphi}{\partial t} dV - \int_{V(t)} \left(\vec{\nabla} \bullet \varphi \vec{v}\right) dV + \int_{V(t)} \vec{\nabla} \bullet \bar{\bar{J}}\, dV - \int_{V(t)} S_g dV = 0 \quad (18)$$

Where φ is a vector containing the variables which are conserved, for example, mass, momentum and energy, per unit of volume. $\bar{\bar{J}}$ is the flow of property per unit of area and time across the surface that bounds the material volume $V(t)$, and S_g is the generation of the property φ per unit of volume and time.

This equation can be rewritten in terms of a partial derivative [15]:

$$\frac{\partial \varphi}{\partial t} + \vec{\nabla} \bullet \varphi \vec{v} + \vec{\nabla} \bullet \bar{\bar{J}} = S_g \quad (19)$$

The first term represents the time rate of change of the property φ, per unit of volume, the second term is the rate of convection per unit of volume, the third term is the surface flux and the fourth term is the volume source.

Therefore the local continuity, momentum and energy equations can be summarised using the following vectors. The first element of the vector φ is the density (mass per unit of volume), the second row is the momentum and the third element is the energy.

$$\varphi = \begin{bmatrix} \rho \\ \rho \vec{v} \\ \rho\left(v + \frac{1}{2}\vec{v}\vec{v}\right) \end{bmatrix}; J = \begin{bmatrix} 0 \\ P\bar{\bar{I}} - \bar{\bar{\sigma}} \\ \vec{q}' - \bar{\bar{T}}\vec{v} \end{bmatrix}; S = \begin{bmatrix} 0 \\ \rho \vec{g} \\ \rho g \vec{v} + \dot{Q}_g \end{bmatrix} \quad (20)$$

The stress tensor $-\overline{\overline{T}}$ represents the normal and shear stresses acting on the surface $\overline{\overline{J}} = -\overline{\overline{T}} = -\left(-p\overline{\overline{I}} + \overline{\overline{\sigma}}\right)$.

These are the instantaneous conservation equations for a single-phase flow. The total number of unknowns is six: velocity vectors (in three directions, thus three unknowns), pressure, temperature and density. An equation of state is also used to close the system of equations.

However, if the flow receives enough heat, it can undergo a phase change. For example, in a CANDU fuel channel, the liquid coolant can change to vapour as it removes the heat generated in the nuclear reactor. In this situation, the flow is said to be a two-phase flow.

4.2 Two-phase flow

There are several definitions of two-phase flow in literature. Yet, one of the most practical is provided by Shire (as cited by Butterwort [8]).

"Two-phase flow is a term covering the interaction flow of two phases (gas, liquid or solid) where the interface between the phases is influenced by their motion. The proviso concerning the interface is inserted to distinguish between problems which are usually considered as two phases and those which are normally accepted as a single phase."

There are more definitions of two-phase systems that explain the difference between multi-component and multi-field systems [14]. This chapter is concerned only with water and vapour two-phase flows.

An important area of nuclear thermal-hydraulics is the two-phase system analysis. Furthermore, under certain specific conditions, such as in a loss of coolant accident (LOCA), additional components such as non-condensable gases can be part of the system, thus adding one component more to the liquid-vapour mixture. In this case, the system is said to be multi-phase and multi-component. The system can be further classified, namely, if the system presents different structural formations, such as liquid droplets or mist, it is also classified as multi-field.

The general local conservation equation takes the following form [15]:

$$\sum_{k=1}^{2}\int_{V(t)}\left(\frac{\partial\varphi_k}{\partial t} - \left(\vec{\nabla}\bullet\varphi_k\vec{v}_k\right) + \vec{\nabla}\bullet\overline{\overline{J}}_k - S_{gk}\right)dV = \int_{A_i(t)}\sum_{k=1}^{2}\left(\vec{u_k}\bullet\varphi\left(\vec{v_k} - \vec{v_i}\right) + \vec{u}_k\bullet\overline{\overline{j}}_k\right)dA \tag{21}$$

$$\varphi = \begin{bmatrix} \rho \\ \rho\vec{v} \\ \rho\left(v + \frac{1}{2}\vec{v}\vec{v}\right) \end{bmatrix}; J = \begin{bmatrix} 0 \\ P\overline{\overline{I}} - \overline{\overline{\sigma}} \\ \vec{q}' - \overline{\overline{T}}\vec{v} \end{bmatrix}; S = \begin{bmatrix} 0 \\ \rho\vec{g} \\ \rho g\vec{v} + \dot{Q}_g \end{bmatrix} \tag{22}$$

Where the subindex k represents a phase, gas (g) or liquid (l).

Note that the first term of Eq. (21) is similar to the single-phase general conservation equation (Eq. (19)). However, the introduction of the summation means that there are two terms, one for the liquid phase and another for the gas phase. Similar to Eq. (19), each of these terms represents the conservation of a property φ. The right side term of Eq. (21) is the interface condition, that is, the properties that are transported from one phase to another via an interface.

Note that there are two velocities for each phase. Under two-phase flow conditions, the lighter phase tends to travel faster than the heavier phase. Similarly, the two-phase mixture can also be in thermodynamic non-equilibrium, that is, each phase can have independent pressure and temperature.

Solving these systems of equations is a very complex task, and is usually prohibitive. Fortunately, this approach is not needed, and the prediction of averaged quantities is usually sufficient.

Averaging procedures have been proposed by numerous researchers, such as Delhaye et al. [15] and Ishii and Hibiki [16].

4.3 Practical two-phase models

As stated in the previous sections, two-phase flow systems are defined by the presence of an interface separating the phases. A more complete description of a two-phase flow system can be obtained using the two-fluid model, which requires solving the conservation equations of mass, momentum and energy for each phase. This system of equations is complex due to the non-linearity nature of fluid dynamics.

To alleviate this complexity, some assumptions have been made to decrease the number of variables, and therefore the number of equations. However, the assumptions made should be in accordance with the nature of the problem. Fortunately, in a nuclear power plant, as in many other plants, the fluid is flowing through pipes where the fluid is essentially axially predominant, resulting in a one-dimensional (1-D) equations. This significantly reduces the complexity of the equations, as these can be written assuming 1-D flow. Nevertheless, there are components or systems where 2-D or 3-D effects should be taken into account, such as the detailed thermal-hydraulics analysis of a fuel bundle.

Some of the most common two-phase models are described in detail in several two-phase flow textbooks [2, 7, 8, 10–17]. For that reason, only the conservation equations are herein presented. The most common models are:

1. Homogeneous equilibrium model;

2. Simplified two-fluid model for separated flow;

3. Drift flux model.

4.3.1 Homogeneous equilibrium model

The Homogeneous Equilibrium Model (HEM) is the simplest model. It assumes thermo-equilibrium between phases, as well as equal velocities between them. That is, the conservation equations used for single-phase flow are also applicable in this model, however, the density (ρ_m) and enthalpy (h_m) are mixture properties.

This model is applicable if one of the phases is finely dispersed, and in which the momentum and energy transfers are sufficiently rapid for the average velocities and the average temperatures of the two phases to be equal.

The conservation equations for the HEM are:

a. Mass conservation equation

$$\frac{\partial \rho_m}{\partial t} + \frac{\partial}{\partial z} \rho_m v = 0 \tag{23}$$

b. Momentum conservation equation

$$\frac{\partial}{\partial t}\rho_m v + \frac{\partial}{\partial z}\left(\rho_m v^2\right) + \frac{\partial P}{\partial z} + \rho_m g = -\frac{4\tau_w}{D} \tag{24}$$

c. Total energy conservation equation (enthalpy)

$$\frac{\partial}{\partial t}\rho_m\left[h_m + \frac{G^2}{2\rho_m^2} + gz\right] - \frac{\partial P}{\partial t} + \frac{\partial}{\partial z}\left[G\left(h_m + \frac{G^2}{2\rho_m^2} + gz\right)\right] = q'''_w \tag{25}$$

4.3.2 Simplified two-fluid model: Separated flow

This is a simple case of a two-fluid model. In this model, it is assumed that the gas and liquid phases are separate, such as in an annular flow as shown in **Figure 5**. The important assumptions for this model are:

- The space and time correlation coefficients are equal to 1.
- The pressure is constant over the cross-sectional area.
- The equation of state valid for local quantities applies to averaged quantities.
- Longitudinal conduction terms in each phase, as well as their derivatives, are negligible.

Applying these assumptions, the conservation equations for this model are:

a. Mass conservation:

$$\frac{\partial}{\partial t}A\alpha\rho_g + \frac{\partial}{\partial z}\left(A\alpha\rho_g v_g\right) = \frac{1}{dz}\delta w \tag{26}$$

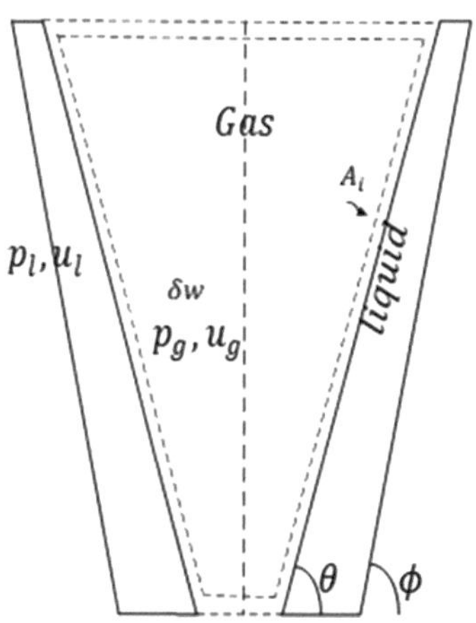

Figure 5.
Control volume for the separate flow model.

$$\frac{\partial}{\partial t}A(1-\alpha)\rho_l + \frac{\partial}{\partial z}(A(1-\alpha)\rho_l v_l) = -\frac{1}{dz}\delta w \tag{27}$$

b. Momentum conservation equation:

$$\frac{\partial}{\partial t}A\alpha\rho_g v_g + \frac{\partial}{\partial z}A\alpha\rho_g v_g^2 + A\alpha\frac{\partial P}{\partial z} + A\alpha\rho_g g = \frac{\delta v}{dz}v_{g_i} - \frac{1}{dz}A_i\, sin\ \gamma\tau_g \tag{28}$$

$$\begin{aligned}&\frac{\partial}{\partial t}A(1-\alpha)\rho_l v_l + \frac{\partial}{\partial z}A(1-\alpha)\rho_l v_l^2 + A(1-\alpha)\frac{\partial P}{\partial z} + A(1-\alpha)\rho_l g\\ &= -\frac{\delta v}{dz}v_{l_i} + \frac{1}{dz}A_i\, sin\ \gamma\tau_g - \frac{1}{dz}A_w\, sin\ \theta\tau_w\end{aligned} \tag{29}$$

$$\frac{\partial}{\partial t}\left[\alpha\rho_g v_g + (1-\alpha)\rho_l v_l\right] + \frac{1}{A}\frac{\partial}{\partial z}A\left[\alpha\rho_g v_g^2 + (1-\alpha)\rho_l v_l^2\right] + \frac{\partial P}{\partial z} + \rho_m g = -\frac{A_w\, sin\ \theta\tau_w}{V} \tag{30}$$

c. Total energy conservation equation (enthalpy):

$$\begin{aligned}\frac{\partial}{\partial t}A\alpha\rho_g h_g^{o-}\frac{\partial}{\partial t}A\alpha P + \frac{\partial}{\partial z}A\ \alpha\rho_g h_g^o v_g = -\frac{1}{dz}\int_{A_i} m_g h_g^o dA + \frac{1}{dz}\int_{A_i} m_g\frac{P}{\rho_g}dA - \frac{1}{dz}\int_{A_i} P\, v_g\left(\vec{u}_g\bullet\vec{u}_z\right)dA\\ +\frac{1}{dz}\int_{A_i}\vec{u}_z\cdot\left(\vec{u}_g\cdot = \sigma_g\right)v_g dA - \frac{1}{dz}\int_{A_i}\left(\vec{u}_g\bullet\vec{q}_g^{\,\prime}\right)dA - \frac{1}{dz}\int_{A_{gw}}\vec{u}_{gw}\cdot\vec{q}_l^{\,\prime}\, dA\end{aligned} \tag{31}$$

$$\begin{aligned}\frac{\partial}{\partial t}A(1-\alpha)\rho_l h_l^{o-}\frac{\partial}{\partial t}A(1-\alpha)P + \frac{\partial}{\partial z}A\ (1-\alpha)\rho_l h_l^o v_l = -\frac{1}{dz}\int_{A_i} m_l h_l^o dA + \frac{1}{dz}\int_{A_i} m_l\frac{P}{\rho_l}dA\\ -\frac{1}{dz}\int_{A_i} P\, v_l\left(\vec{u}_l\bullet\vec{u}_z\right)dA + \frac{1}{dz}\int_{A_i}\vec{u}_z\cdot\left(\vec{u}_l\cdot = \sigma_l\right)v_l dA - \frac{1}{dz}\int_{A_i}\left(\vec{u}_l\bullet\vec{q}_l^{\,\prime}\right)dA - \frac{1}{dz}\int_{A_w}\vec{u}_{lw}\cdot\vec{q}_l^{\,\prime}\, dA\end{aligned} \tag{32}$$

$$\frac{\partial}{\partial t}A\left[\alpha\rho_g\left(h_l^o - \frac{P}{\rho_g}\right) + (1-\alpha)\rho_l\left(h_g^o - \frac{P}{\rho_l}\right)\right] + \frac{\partial}{\partial z}A\ \left[\ \alpha\rho_g h_g^o v_g + (1-\alpha)\rho_l h_l^o v_l\right] = q'_w \tag{33}$$

Where

$$h^o = h_k + \frac{1}{2}v_k^2 + gz \tag{34}$$

The subindexes g and l represent the gas and liquid phases, the subindex i represents the property evaluated at the interface, A is the area, m is the average mass transfer per unit of interface area, P is the average pressure, q is heat, u is the velocity at the interface, w is property evaluated at the wall, $\overline{\overline{\sigma}}$ and τ are wall stresses.

The reader is recommended to refer to reference [17] for further information on this model.

4.3.3 Drift flux model

In earlier two-phase flow models, the fluid was treated as a homogeneous mixture of liquid and vapour, and consequently, only three conservation equations were needed to describe the two-phase flow. In the HEM, the phases are assumed to move at the same velocity, while also experiencing the same temperature. An extension to the mixture model is the drift flux model, in which the relation between the phasic velocities is described through an algebraic equation, thus allowing for a slip between the phases. The Zuber and Findlay [19] drift flux model is probably the most common model that takes into account the relative velocity of the phases. This model uses a drift velocity v_{gj} and a void distribution parameter C_o. This model uses the following relationships:

$$j_l = (1 - \alpha)v_l \tag{35}$$

$$j_g = \alpha v_g \tag{36}$$

$$j = j_l + j_g \tag{37}$$

$$v_{gj} = v_g - j = (1 - \alpha)v_g - v_l \tag{38}$$

Where α denotes the gas local time fraction, that is, the local void fraction, and v_k is the component, along the axis of the pipe, of the local time-averaged velocity.

We have locally:

$$\alpha v_{gj} = \alpha v_g - \alpha j = \alpha_j - \alpha j \tag{39}$$

Averaging this equation over the total cross-section of the pipe, we obtain the Zuber and Findlay void equation:

$$\langle\langle \alpha \rangle\rangle = \frac{\langle\langle j_g \rangle\rangle}{C_o \langle\langle j \rangle\rangle + \tilde{v}_{gj}} = \frac{J_g}{C_o J + \tilde{v}_{gj}} \tag{40}$$

Where J_g and J are the gas and mixture superficial velocities. In this equation, two quantities appear: (1) The C_o parameter, which accounts for the shape of the α and j profiles, and (2) the void weighted average of the local drift velocity $\tilde{v}_{gj}$, which takes into account the relative velocities between the phases. The symbol $\langle\langle\rangle\rangle$ defines the average value of a variable x over the cross-sectional area A as

$$\langle\langle x \rangle\rangle = \frac{1}{A} \int_A x \, dA \tag{41}$$

To calculate $\langle\langle \alpha \rangle\rangle$ from this equation is sufficient to know J_g, J, that is, the volumetric flow rates of each phase and the cross-sectional area, the distribution parameter C_o.

$$C_o = \frac{\langle\langle \alpha j \rangle\rangle}{\langle\langle \alpha \rangle\rangle \langle\langle j \rangle\rangle} \tag{42}$$

and the local drift velocity $\tilde{v}_{gj}$:

$$\tilde{v}_{gj} = \frac{\langle\langle \alpha v_{gj} \rangle\rangle}{\langle\langle \alpha \rangle\rangle} \tag{43}$$

Zuber and Findlay [19] noticed that C_o and $\tilde{v}_{gj}$ are only functions of the flow regime. The authors, therefore, recommended certain values presented in the next table. Nowadays, there are several expressions for $\tilde{v}_{gj}$ and C_o available in scientific literature [20].

Note this drift flux model can be applied to homogeneous two-phase flow models to take into account the relative velocities between the gas and liquid phases. However, care must be taken to ensure the distribution parameter C_o and drift velocity $\tilde{v}_{gj}$ are applicable to the type of problem and flow regime.

Some models for distribution parameter C_o and drift velocity $\tilde{v}_{gj}$ are presented in **Table 1**.

Flow regime	Distribution parameter C_o	Drift velocity
Bubbly	• Circular cross-section: *Reduced Pressure* $\equiv P_R = \frac{P}{P_c}$ • $D > 5\,cm\ C_o = 1 - 0.5P_R$ • $D < 5\,cm$ • $P_R < 0.5\ C_o = 1.2$ • $P_R > 0.5\ C_o = 1.4 - 0.4P_R$	$\tilde{v}_{gj} = 1.41\left(\frac{\sigma g \Delta\rho}{\rho_l^2}\right)^{\frac{1}{4}}$
Slug	$C_o = 1.2$	$\tilde{v}_{gj} = 0.35\left(\frac{g\Delta\rho D}{\rho_l}\right)^{1/4}$
Annular	$C_o = 1.0$	$\tilde{v}_{gj} = 23\left(\frac{\mu_l j_l}{\rho g D}\right)^{\frac{1}{2}}\left(\frac{\Delta\rho}{\rho_l}\right)$

Table 1.
Distribution parameter C_o and drift velocity $\tilde{v}_{gj}$ [15].

Model designation	Restriction		No. of field equations			No. of interface transfer equations	External constitutive equations	
	No	Imposed on	Mass	Momentum	Energy		No	Type
1V1T	3	$v_g = v_l, h_l, h_G$	1	1	1	0	2	$\tau, \overline{q}$
1VS1T		$\frac{v_g}{v_l}, h_l, h_G$					3	$\tau, \overline{q}, \frac{v_g}{v_l} = slip$
1VD1T		v_g, v_l, h_l, h_G					3	$\tau, \overline{q}, v_r = v_g - v_l$ or $v_{gm} = V_g - v_m$ or $v_{gj} = v_g - v_j$
$1VT_KT_{SAT}$	2	$v_g = v_l, h_l$	1	1	2	1	3	$\tau, \overline{q}, q_g, E$
$1VST_KT_{SAT}$		$\frac{v_g}{v_l}$	1	1	2		4	$\tau, \overline{q}, q_g, \frac{v_g}{v_l}, E$
$1VDT_KT_{SAT}$		$v_g - v_l, h_G$	2	1	1		4	$\tau, \overline{q}, \Gamma, v_r\,(or\ v_{gm} or\ v_{gj})$
2V1T		h_l, h_G	1	2	1		4	$\tau_l, \tau_g, \overline{q}, M$
1V2T	1	$v_g = v_l$	2	1	2	2	5	$\tau_l, q''_l, q''_g, \Gamma, E$
1VD2T		$v_g - v_l$	2	1	2		6	$\tau_l, q''_l, q''_g, \Gamma, E, v_r\,(or v_{gm})$
$2VT_KT_{SAT}$		h_l *or* h_G	2	2	1		5	$\tau_l, \tau_g, \overline{q}, q_k, M$
$2VT_KT_{SAT}$		h_l *or* h_G	1	2	2		6	$\tau_l, \tau_g, \overline{q}, q_k, M, E$
2V2T	0	None	2	2	2	3	7	$\tau_l, \tau_g, q_l, q_g, \Gamma, M, E$

Table 2.
Practical two-phase flow models [15].

Notice that one of the most important characteristics of multi-phase flow is that it can take on different spatial distributions. This characteristic poses a major challenge to an exact solution treatment [17]. In fact, the majority of the practical two-phase models for conduits, such as the ones presented in this section, were developed for a specific flow regime(s). The following section describes in greater detail the flow regimes for vertical and horizontal flows. **Table 2** presents different two-phase flow models based on different assumptions and simplifications [15].

5. Flow regimes

The hypothetical experiment used to explain the boiling curve in the previous subsection describes some of the different regimes a flow can experience during a boiling process. For example, if the coolant undergoes a boiling process, the liquid-gas mixture will present different field structures that give a macroscopic behaviour to the mixture. These structures, also known as regimes or patterns, are classified by visual observation. For that reason, there is no unique set of flow regimes. In thermal-hydraulics analysis, the determination of the flow regime is key, as several variables are dependent on the flow regime, such as the heat transfer coefficient.

In the case of a gas-liquid mixture, gravity plays an important role, as the lighter phase tends to separate from the heavier phase, therefore, the flow regime depends on the orientation of the system. There are different regime classifications for vertical, horizontal and inclined systems.

One classification for vertical and horizontal orientation is presented in **Figure 6** (**a**) and (**b**) respectively.

The best way to understand each regime is to examine the behaviour of the coolant as it flows through a fuel channel, as shown in **Figure 7**. The fuel rod is assumed to produce heat, and the flow is assumed to be forced. Normally, one flow pattern transitions into another when the heat flux is changed, or when the mass flux is raised or lowered. For example, subcooled boiling can easily become saturated boiling, and saturated boiling can immediately become bulk boiling. The transition points between these flow regimes are both mass flux and pressure dependent. They depend on the saturation temperature, as well as the coolant channel geometry. However, for a circular coolant channel, these regimes are relatively simple to understand. A graphical depiction of how these flow regimes interact is called a *flow regime map*.

In some fuel channels of a CANDU-6 reactor, the coolant enters the fuel channel as a subcooled liquid and exits the fuel channel as a two-phase mixture. Under these conditions, the velocities of the liquid and vapour phases are different. Thus, heat transfer coefficients become *flow regime dependent*.

5.1 Vertical flows

Masterson [7] gives a detailed analysis of flow patterns commonly observed in vertical co-current flows, such as those occurring in Pressurised Water Reactor (PWR) and Boiling Water Reactor (BWR) cores.

Regime I: Subcooled Liquid Flow. When a coolant enters the inlet of a reactor fuel assembly, it is normally a *subcooled liquid*. The temperature of the coolant stays below the saturation temperature until it progresses further into the core.

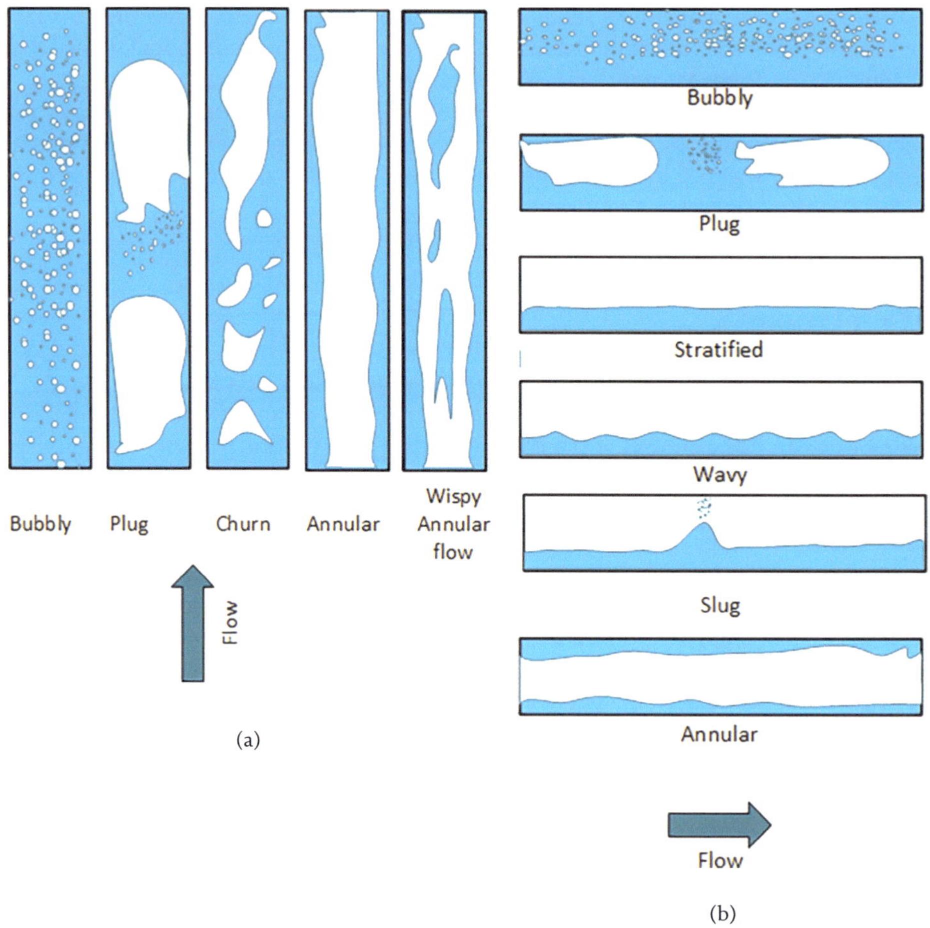

Figure 6.
Flow regimes for (a) vertical (b) horizontal co-current flows.

Regime II: Bubbly Flow with Subcooled Boiling. Once the temperature of the cladding reaches the saturation temperature, small bubbles begin to form on the surface of the rods. These bubbles detach from the surface and flow into the turbulent core where the bulk fluid temperature is at least several degrees below the saturation temperature ($T_{BULK} < T_{SAT}$). Any bubbles that form in this way are immediately engulfed by the cooler liquid, and they collapse back into the flow stream. The creation of vapour bubbles at the wall surface is called *nucleate boiling* and these bubbles do not merge together to form larger bubbles or voids. Hence, the bulk fluid temperature stays below the saturation temperature, and the two-phase mixture tends to be highly turbulent.

Regime III: Bubbly Flow with Saturated Boiling. *Bubbly flow with saturated boiling* is similar to bubbly flow with subcooled boiling, except that the average temperature of the fluid has now reached the saturation temperature. When this occurs, the bubbles that detach from the surface of the cladding do not immediately collapse when they enter the flow stream. Instead, they either remain intact or combine with other bubbles to form larger bubbles or *voids*.

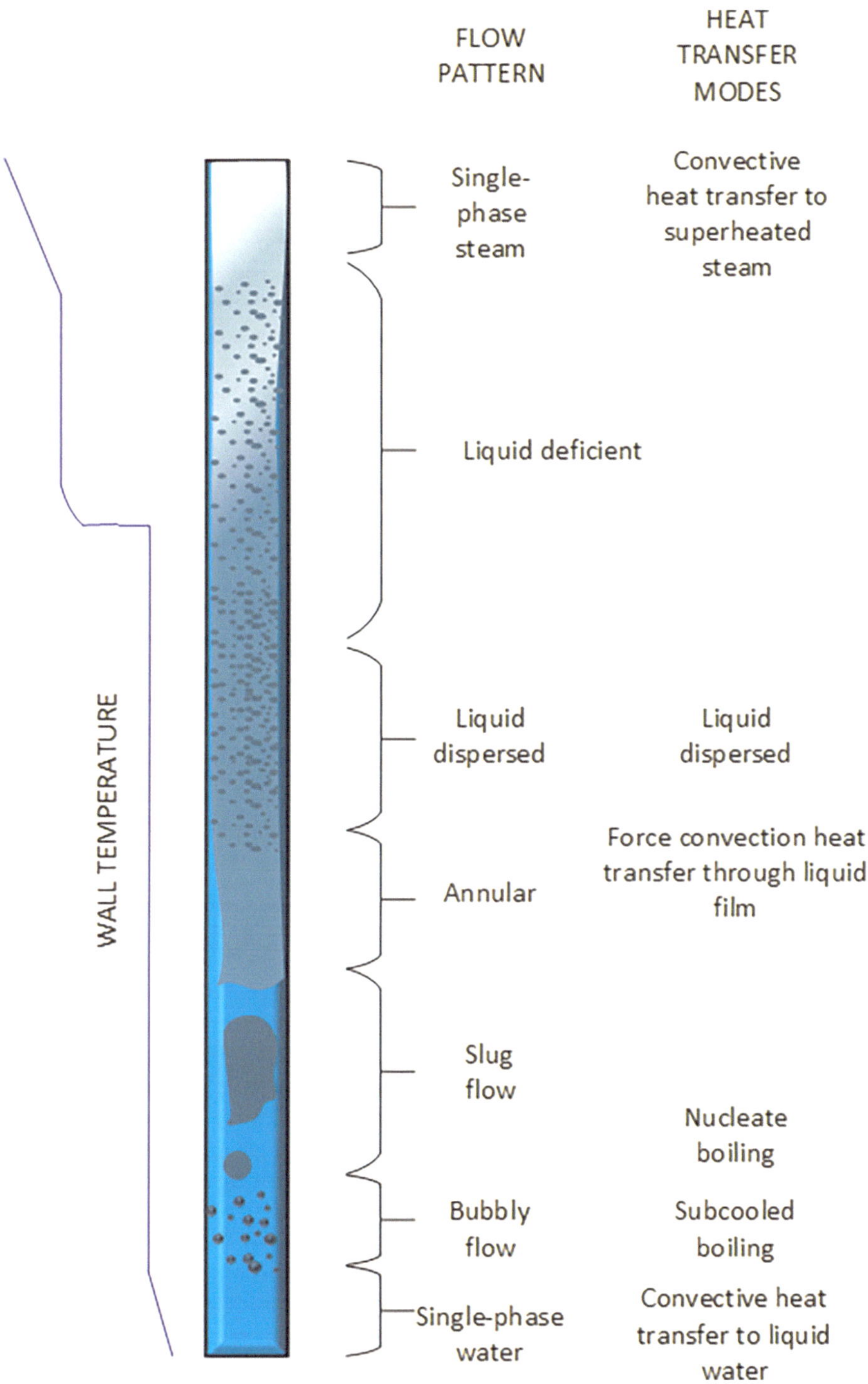

Figure 7.
Flow regimes for a vertical heated channel.

Regime IV: Saturated or Bulk Boiling *Saturated or bulk boiling* is similar to bubbly flow, except that the void fraction now rises from 40 to 60%.

Regime V: Slug Flow. In the *slug flow regime*, the bubbles that have already formed coalesce into very large bubbles that not only span the entire width of the channel but also remain long and continuous. In reactors, these vapour bubbles can be 4–6 times *longer* than they are wide. The liquid film on the surface of the rods cannot be agitated anymore, because there is no additional turbulent mixing to increase the heat transfer rate.

Regime VII: Annular Flow. The primary difference between the *annular flow regime* and the churn flow regime is that *the liquid film on the surface of the fuel rods is now moving in the same direction as the vapour*. This causes the "churn" in the flow pattern to nearly disappear because the interfacial friction between the two phases is now lower (due to a reduction in the relative velocity difference between the phases). Sometimes this interfacial friction is called *interfacial drag*.

Regime IX: Pure Vapour Flow. In this final flow regime, the liquid completely disappears and all that remains is pure vapour flow. Hence, this regime is called the *pure vapour flow* regime. The heat transfer coefficients behave in exactly the same way as they do for single-phase flow. Only in this case, the single phase is the vapour phase.

5.2 Horizontal flows

The flow patterns for co-current horizontal flow are applicable to CANDU reactors. There are several studies that were performed under horizontal and adiabatic (air-water) test sections. However, adiabatic horizontal air-water systems using fuel bundles are limited.

Osamusali et al. [21] presented the generally accepted horizontal flow pattern classifications. These are:

Stratified flow: The stratified flow regime is characterised by the liquid flowing at the bottom of the test section, and the gas phase at the top. This can be further classified: (1) *stratified smooth flow*. In this case, a smooth gas-liquid interface exists. At high-flow rates, the interface may become wavy. In this case, it is referred to as *stratified wavy flow* pattern. In fuel bundles such as those used in CANDU reactors, interfacial waves may result from disturbances at the end plates.

Intermittent flow: The intermittent flow regime is characterised by liquid bridging the gap between the gas-liquid interface and the top of the pipe. The liquid bridges are separated by stratified flow zones. The intermittent flow is subdivided into the *plug flow* regime, occurring at low gas velocities and having a liquid bridge free of gas bubbles, and the *slug flow* regime, which occurs at higher gas-flow rates and entrains a significant amount of gas bubbles in the liquid bridge. During intermittent flows in rod bundles, the liquid bridges across the elements at the upper part of the channel.

Annular Flow: The annular flow pattern is characterised by the liquid phase flowing around the inner periphery of the pipe and surrounding a core of a fast-flowing gas phase. The gas core may entrain some liquid droplets, and the gas-liquid interface is generally wavy. At low gas-flow rates, the liquid essentially flows as a thick film at the bottom of the pipe with rather unstable waves at the gas-liquid interface, continuously swept up around the pipe periphery, resulting in the *wavy-annular flow* regime. This eventually leads to the fully developed *annular flow* regime at higher gas-flow rates, characterised by a continuous liquid film around the inner periphery of the pipe. During annular flows, the rod-bundle elements in the gas core may be covered with very thin liquid films.

Bubbly Flow: The bubbly flow regime is characterised by the void being in the form of discrete bubbles, which are distributed throughout the continuous liquid phase that otherwise fills the pipe section. The bubble concentration is highest at the top of the pipe, especially at lower mass velocities.

Osamusali et al. [21] performed a series of experiments to study the transition of flow patterns in a CANDU-type fuel channel. These authors concluded that the descriptions of two-phase flow patterns occurring in fuel bundles are similar to those observed in pipes. However, Yang [22] observed in a more recent study, that for crept fuel channels the bundle geometry has a strong effect on flow patterns. This effect was pronounced at the transition between stratified and plug/slug flow. For the transition between stratified wavy annular and wavy annular flow, the effect of crept fuel channel was small or non-existent. Bundle misalignment did not show an impact on flow stratification.

5.2.1 Flow regime analysis for CANDU fuel channels

Osamusali et al. [21] developed a generalised flow regime map for a horizontal 37-element bundle, obtained using air-water experimental data at room temperature and near atmospheric pressure. The dimensionless representation is based on dimensionless quantities $X, F, \& T$. Osamusali et al. [21] presented the following model:

$$X = \sqrt{\left(\frac{dP}{dx}\right)_l / \left(\frac{dP}{dx}\right)_g} \tag{44}$$

The parameter X can be expressed as a function of the superficial velocities:

$$X = \left(\frac{\mu_l}{\mu_g}\right)^{0.108} \left(\frac{\rho_l}{\rho_g}\right)^{0.392} \left(\frac{j_l}{j_g}\right)^{0.892} \tag{45}$$

Where j_l and j_g are the superficial velocities of the gas and liquid phases respectively. The superficial velocity represents the case when the gas of the liquid is assumed to be flowing alone in the channel. For fuel bundles, the superficial velocity is based on the channel cross-sectional area excluding the area occupied by the fuel elements. Using a friction factor developed for a bare 37-element fuel bundle, $f = 0.243\mathbf{Re}^{-0.216}$, Eq. (45) can be recast as:

$$X = \left(\frac{\mu_l}{\mu_g}\right)^{0.108} \left(\frac{\rho_l}{\rho_g}\right)^{0.5} \left(\frac{1-x}{x}\right)^{0.892} \tag{46}$$

The dimensionless parameter **F** is the modified Froude number, thus given as:

$$\mathbf{F} = \sqrt{\left(\frac{\rho_g}{\left(\rho_l - \rho_g\right) g D_h}\right)} j_g \tag{47}$$

D_h is the hydraulic diameter, obtained using the channel cross-sectional area excluding the area occupied by the fuel elements, and the total wetted perimeter of the fuel elements and wall channel. F can be expressed in terms of the total mass flux G and flow quality x as

$$\mathbf{F} = \sqrt{\left(\frac{Gx}{\rho_g\left(\rho_l - \rho_g\right)gD_h}\right)} \tag{48}$$

6. Pressure drop models

Pressure drop is calculated by writing the conservation equations of mass, momentum and energy, and then evaluating the pressure difference. However, in order to solve the basic conservation equations in thermal-hydraulic system codes, additional constitutive equations are required, simply because the number of unknowns is higher than the number of equations. Among them, a constitutive equation is needed to take into account the shear stress.

As mentioned in Section 2, the power generated in the reactor core is removed by the coolant mass flow rate. In a CANDU 6, this mass flow is driven by 4 pumps, which have been dimensioned according to pressure drop calculations and other parameters. Therefore, the design of the reactor plant requires knowledge of the pressure losses across the individual components of the plant. Furthermore, the pressure drop is needed to assess various postulated events and to dimension additional system structures and components needed to support the design requirements of the plant.

In a closed conduit, such as a nuclear fuel channel in a CANDU-6, the pressure drop can be calculated by adding individual pressure drop components as follows:

$$\Delta P_{tot} = \Delta P_f + \Delta P_{acc} + \Delta P_{grav} + \Delta P_K \tag{49}$$

where ΔP_{tot} is the total pressure drop, ΔP_{fr} is the pressure due to the frictional resistance, ΔP_{acc} is the pressure drop due to acceleration of the flow, ΔP_{grav} is the pressure drop due to gravity, and ΔP_K is the pressure drop due to local flow obstructions.

6.1 Frictional resistance and friction factor

The frictional resistance is the shear stress between the flow and the contacting wall. To account for this shear stress, a non-dimensional friction factor, or the Darcy-Weisbach equation, is commonly used to interrelate the frictional pressure drop to the wall shear stress [23]:

$$f = \frac{\left(\frac{dP}{dz}\right)D}{\frac{1}{2}\rho u^2} \tag{50}$$

Where the frictional pressure gradient is negative.

Several correlations and models exist for estimating the friction factor. For example, for turbulent flows in a smooth conduit, the Blasius approximation is widely used [24]:

$$f = 0.316\,\mathbf{Re}^{-0.25} \quad for \quad 4\times10^3 < \mathbf{Re} < 1\times10^5 \tag{51}$$

The Filonenko correlation is also recommended [24]:

$$f = \frac{1}{(1.82\log_{10}\mathbf{Re}_b - 1.64)^2} \quad for \quad 4\times10^3 < \mathbf{Re} < 10^{12} \tag{52}$$

However, these correlations only apply to smooth pipes. In reality, most surfaces present some roughness, such as pressure tubes in a CANDU reactor. For that reason, Colebrook combined the smooth wall and fully rough relations in an implicit formula (23):

$$f = \left[-2\log\left(\frac{\epsilon/D}{3.7}\right) + \frac{2.51}{\sqrt{f}\,\mathbf{Re}}\right]^{-2} \quad for \quad \mathbf{Re} > 4\times10^3 \tag{53}$$

However, these models are only valid for isothermal flows or flows without large changes in properties (especially viscosity and density, as these are sensitive to temperature variations).

Furthermore, there are friction factor correlations that were developed for CANDU 37-element fuel bundles [25].

Snoek and Ahmad proposed (as cited in [25]):

$$f = 0.050\,\mathbf{Re}^{-0.057} \quad for \quad 108\,000 \le \mathbf{Re} \le 418\,000 \tag{54}$$

Venkat Raj (As cited in [25]) proposed the following correlations for a horizontal 37-element bundle with split-wart spacers, and included the effect of the junctions:

$$f = 0.22\mathbf{Re}^{-0.163} for\ 10\,000 \le \mathbf{Re} \le 140\,000 \tag{55}$$

$$f = 0.108\mathbf{Re}^{-0.108} for\ 140\,000 \le \mathbf{Re} \le 50\,000 \tag{56}$$

To take into account the non-isothermal (*non-iso*) nature of these types of flows in calculating the frictional pressure loss, the common approach consists of introducing correction factors into an isothermal friction factor correlation (f_{iso}) These correction factors usually take the form of ratios between a fluid bulk property (such as viscosity or density) and the property evaluated at the wall conditions. For example, the following correlations are generally used for supercritical water conditions:

The Kirillov correlation (as cited in [24]):

$$f_{non-iso} = f_{iso}\left(\frac{\rho_w}{\rho_b}\right)^{0.4} \tag{57}$$

Leung and Groeneveld proposed (as cited in [25]):

$$f_{non-iso} = f_{iso}\left(\frac{\mu_b}{\mu_w}\right)^{-0.28} \tag{58}$$

6.2 Acceleration

For a one-dimensional flow in a conduit with axial density variation, the pressure drop due to acceleration can be calculated as [24]:

$$\Delta P_{acc} = G^2\left(\frac{1}{\rho_{out}} - \frac{1}{\rho_{in}}\right) \tag{59}$$

6.3 Gravity

This pressure drop component takes into account the gravity force acting on the mass flow. This term, usually called static head, is computed as [24]:

$$\Delta P_{grav} = \overline{\rho} g L \ \sin\theta \tag{60}$$

Where $\overline{\rho}$ is the average density between the inlet and outlet test section, L is the length of the test section, and θ is the inclination angle.

6.4 Flow obstructions

This component takes into account the obstacles the fluid experiences in a flow, such as endplate spacers in a CANDU fuel channel. This component is generally computed as [24]:

$$\Delta P_K = K\frac{1}{2}\rho u^2 = K\frac{G^2}{2\rho} \tag{61}$$

Where K is a form loss coefficient usually obtained from experimental data or correlations.

6.5 Two-phase pressure drop

In CANDU-6 reactors the flow is allowed to change phase at the end of some fuel channels. The maximum allowed thermodynamic quality is 4%. Therefore some fuel channels experience two-phase flow conditions. In addition, there are several postulated accidents where two-phase flow conditions are predominant, such as LOCA events. The pressure drop under these two-phase conditions differs greatly from single-phase flow conditions, because the lighter phase tends to travel faster than the heavier phase to satisfy the mass conservation, adding (1) the pressure drop component called acceleration pressure dropΔP_{acc}, and (2) the pressure losses between the flow and the wall is affected by the presence of a second phase and the flow pattern. In diabatic cases, such as a fuel channel, the heated surface of the element changes the properties near the heated wall, which in turn changes the boundary layer, and thus affecting the friction factor. There are several empirical models to predict the pressure drop under two-phase conditions. Most of them are based on the homogeneous and the separated-flow models (see previous Section 4.3). Most commonly, empirical correlations are used to determine a two-phase friction multiplier. There are several models available, and these vary according to the degree of complexity and might depend on the flow regime.

The two-phase multiplier approach, which is the basis for most of the cited methods, is the generally accepted engineering model to account for the effect of a two-phase mixture in a flow channel. The idea behind this approach is to calculate the pressure drop of one phase (gas or liquid) ΔP_L first. To determine the two-phase pressure drop ΔP_{TP}, the single-phase pressure drop is multiplied with a two-phase multiplier $\Phi_{2_{l_g}}$ to consider the influence of the second phase. The following four basis methods to define the two-phase multiplier are commonly used: – two methods that assume the liquid or gas phase flowing alone in the flow channel—two methods that assume the entire mixture flowing as liquid or gas only. Applying these basis definitions of the two-phase multipliers, the two-phase friction pressure gradients can be expressed as follows [26]:

$$\left\{\frac{\Delta P}{\Delta z}\right\}_{TPF} = \left\{\frac{\Delta P}{\Delta z}\right\}\Phi_l^2 \tag{62}$$

$$\left\{\frac{\Delta P}{\Delta z}\right\}_{TPF} = \left\{\frac{\Delta P}{\Delta z}\right\}\Phi_g^2 \tag{63}$$

$$\left\{\frac{\Delta P}{\Delta z}\right\}_{TPF} = \left\{\frac{\Delta P}{\Delta z}\right\}\Phi_{lo}^2 \tag{64}$$

$$\left\{\frac{\Delta P}{\Delta z}\right\}_{TPF} = \left\{\frac{\Delta P}{\Delta z}\right\}\Phi_{go}^2 \tag{65}$$

The $\Phi_L^2, \Phi_G^2 \Phi_{LO}^2, \Phi_{GO}^2$terms constitute the two-phase pressure drop multiplier to be determined.

6.5.1 Modifications of the single-phase friction factor based on the homogeneous model

In this case, the flow is treated as a pseudo-single-phase fluid, and the friction pressure drop is calculated using a modified friction factor, f_{TP}. In this case, the shear stress at the wall is assumed as:

$$\Delta P_{TP} = f_{TP}\frac{L}{D_h}\left(\frac{u}{2\rho_m A^2}\right) \tag{66}$$

Where the f_{TP} is the two-phase friction factor, $\overline{\rho}$ is the mixture density. The friction factor is calculated using friction factor correlations such as Eqs. (53)–(56).

To calculate the **Re** of the flow, mixture properties are used. For example, the mixture viscosity can be estimated as:

$$\frac{1}{\mu} = \frac{x}{\mu_g} + \frac{1-x}{\mu_l} \tag{67}$$

or

$$\mu = x\mu_g + (1-x)\mu_l \tag{68}$$

Empirical calculations applying the two-phase multiplier concept based on the separated flow model.

The frictional pressure drop in two-phase flow is generally based on separate flow models, such as the one presented in Section 4.3.2. Recall that in the separated flow model the phases are considered to flow separately in a flow channel, as each phase has its own velocity.

Martinelli and Nelson [27] and Lockhart and Martinelli [28] developed the base for the two-phase friction multiplier approach. Lockhart and Martinelli [28] defined a parameter χ, known as the Martinelli parameter:

$$\chi^2 = \frac{\Phi_g^2}{\Phi_l^2} = \frac{(dP/dz)_l}{(dP/dz)_g} \tag{69}$$

The study of two-phase pressure drop has been subject to numerous investigations over the last six decades. Most of the developed methods are based on empirical models.

Finally, the reader is referred to a technical book prepared by an IAEA Coordinated Research Project (CRP), which contains several models for predicting heat transfer coefficients and pressure drop in different flow regimes [25].

Nomenclature

A	coefficient $\frac{3.63P}{VE_R\Sigma_f}$ (Neutrons/m^{2}s)
C_p	Specific heat capacity (J/kg°C)
C_o	distribution parameter (–)
D	diameter (m)
E	energy transfer between phases (J)
E_d	energy deposited locally in the fuel per fission (J)
E_R	recoverable energy (J/fission)
f	friction factor (—)
G	mass flux (kg/m^{2}s)
g	acceleration of gravity (m/s^2)
h	specific enthalpy (J/kg)
$\tilde{h}$	heat transfer coefficient (W/m^{2}°C)
H_e	height extrapolated length (m)
$\bar{\bar{I}}$	irreversibility or lost work (J/s)
$\bar{\bar{J}}$	generalised surface source or sink for mass, momentum and energy (—)
j	superficial velocity (m/s)
J_o	Bessel function of order zero (—)
K	form loss coefficient (—)
k	thermal conductivity (W/m°C)
L	test section length (m)
M	momentum transfer between the phases (kg/m s)
P	pressure (pa)
$\dot{Q}_g$	power (W)
q',	linear heat generation rate (W/m)
q'',	heat flux (W/m^2)
q'''	volumetric heat generation rate (W/m^3)
$\bar{q}$	heat flux between the two-phase and the mixture and the wall (W/m^2)
r	radius coordinate (—)
R_e	radius extrapolated length (m)
s_g	generation of a property per unit of volume and time
T	temperature (°C)
t	time (s)
u	interfacial velocity (m/s)
v	velocity (m/s)

V	volume (m^3)
v_{gj}	drift velocity (m/s)
X	dimensionless parameter (—)
x	quality (—)
z	axial coordinate (—)

Greek letters

α	void fraction (—)
Γ	net vapour volumetric production rate resulting from phase change (kg/m^3 s)
γ	angle (degrees)
δw	phase evapouration rate (kg/s)
ϵ	roughness (m)
Θ	neutron flux (Neutrons/m^2s)
θ	angle (rad)
χ^2	two-phase pressure drop parameter "Martinelli" (—)
Φ^2	two-phase pressure drop multiplier (—)
μ	viscosity (kg/m s)
φ	property per unit of volume ($1/m^3$)
ρ	density (kg/m^3)
Σ_f	macroscopic fission cross-section (cm^{-1})
$\overline{\overline{\sigma}}$	normal stress component (N/m^2)
τ	shear stress component (N/m^2)

Non-dimensional numbers

F	modified Froude number $\mathbf{F} \equiv \sqrt{\left(\frac{\rho_g}{(\rho_l-\rho_g)gD_h}\right)} j_g$ (—)
Nu	Nusselt number, $\equiv \frac{hD}{k}$ (—)
Pr	Prandtl number, $\equiv \frac{C_p\mu}{k}$ (—)
Re	Reynolds number, $\equiv \frac{GD_e}{\mu}$ (—)
$\boldsymbol{P_r}$	reduced pressure, $\equiv \frac{P}{P_c}$ (—)

Mathematical symbols

$\langle\langle\rangle\rangle$	Cross section average (—)
∇	Del (also known as Nabla) operator (—)
$\rightarrow$	vector (—)

Subscripts or superscripts

acc	acceleration
b	bulk
c	critical

centre	property evaluated at the centre
e	extrapolated length
f	friction factor
g	gas phase
grav	gravity
in	inlet
iso	isothermal
K	form loss
k	phase (*g*=gas, *l*=liquid)
l	liquid phase
non-iso	non-isothermal
out	outlet
R	reduced
tot	total
TPF	two-phase flow
w	wall

Acronyms and abbreviations widely used in text and list of references

CANDU	CANada Deuterium Uranium
CRP	Coordinated Research Project
D_2O	deuterium
DNB	Departure Nucleate Boiling
EFP	End-Flux-Peaking
HEM	Homogeneous Equilibrium Model
IAEA	International Atomic Energy Agency
LOCA	loss-of-coolant accident
NPP	Nuclear Power Plant

Author details

Armando Nava Dominguez
Canadian Nuclear Laboratories, Chalk River, Canada

*Address all correspondence to: armando.nava@cnl.ca

References

[1] Lamarsh JR, Baratta AJ. Introduction to Nuclear Engineering. Prentice Hall; 2001. Available from: https://books.google.ca/books?id=xSNnQgAACAAJ

[2] Duderstadt JJ, Hamilton LJ. Nuclear Reactor Analysis. New York: Wiley; 1976

[3] International Atomic Energy Agency, Thermophysical Properties Database of Materials for Light Water Reactors and Heavy Water Reactors IAEA-TECDOC-1496

[4] Glasstone S, Sesonske A. Nuclear Reactor Engineering: Reactor Systems Engineering. US: Springer; 2011. Available from: https://books.google.ca/books?id=iSizoAEACAAJ

[5] Dittus FW, Boelter LMK. Heat transfer in automobile radiators of the tubular type. International Communications in Heat and Mass Transfer. 1985;**12**:3-22. DOI: 10.1016/0735-1933(85)90003-X

[6] Sieder EN, Tate GE. Heat transfer and pressure drop of liquids in tubes. Industrial and Engineering Chemistry. 1936;**28**:1429-1435. DOI: 10.1021/ie50324a027

[7] Masterson R. Nuclear Reactor Thermal Hydraulics: An Introduction to Nuclear Heat Transfer and Fluid Flow. CRC Press; 2019. Available from: https://books.google.com.co/books?id=_2ytDwAAQBAJ&dq=heat+transfer+convection+coefficient+pool+boiling+of+water+1200&hl=es&source=gbs_navlinks_s

[8] Butterworth D, Hewitt GF. Two-Phase Flow and Heat Transfer. Oxford University Press; 1979. Available from: https://books.google.ca/books?id=EyKhPwAACAAJ

[9] Bejan. Convection Heat Transfer. Wiley; 2013. Available from: https://books.google.ca/books?id=9yC91-gpU8sC

[10] Bergles AE, Collier JG, Mayinger F. In: Bergles AE et al., editors. Two-phase Flow and Heat Transfer in the Power and Process Industries. Hemisphere Publishing Corporation; 1981. Available from: https://books.google.ca/books?id=Zg8pAQAAMAAJ

[11] Todreas NE, Kazimi MS. Nuclear Systems: Thermal Hydraulic Fundamentals. Hemisphere Pub. Corp.; 1990. Available from: https://books.google.ca/books?id=j2xTAAAAMAAJ

[12] Todreas NE, Kazimi MS. Nuclear Systems. Taylor & Francis; 1990. Available from: https://books.google.ca/books?id=zwyToqVDVGwC

[13] Hsu YY, Graham RW. Transport Processes in Boiling and Two-Phase Systems, Including near-Critical Fluids. American Nuclear Society; 1986. Available from: https://books.google.ca/books?id=IQ1RAAAAMAAJ

[14] Kolev NI. Multiphase Flow Dynamics 1: Fundamentals. Berlin, Heidelberg: Springer - Verlag; 2015. DOI: 10.1007/978-3-319-15296-7

[15] Delhaye JM, Giot M, Riethmuller ML. Thermohydraulics of Two-Phase Systems for Industrial Design and Nuclear Engineering. Hemisphere Publishing Corporation; 1980. Available from: https://books.google.ca/books?id=k2tTAAAAMAAJ

[16] Ishii M, Hibiki T. Thermo-fluid dynamics of two-phase flow. In: Thermo-Fluid Dyn. Two-Phase Flow. 2nd ed. New York: Springer; 2011. p. 518. DOI: 10.1007/978-1-4419-7985-8_1

[17] Lahey RT, Moody FJ. American Nuclear Society, Research USE, Administration D. The Thermal Hydraulics of a Boiling Water Nuclear Reactor. La Grange Park, Illinois, USA: The Society; 1977

[18] Tong LS, Tang YS. Boiling Heat Transfer and Two-Phase Flow. Taylor & Francis; 1997. Available from: https://books.google.ca/books?id=xNGhaNouY6oC

[19] Zuber N, Findlay JA. Average volumetric concentration in two-phase flow systems. Journal of Heat Transfer. 1965;**87**:453-468. DOI: 10.1115/1.3689137

[20] Hammouda N, Rao YF. The effect of the advanced drift-flux model of ASSERT-PV on critical heat flux, flow and void distributions in CANDU bundle subchannels. Annals of Nuclear Energy. 2017;**104**:197-213. DOI: 10.1016/j.anucene.2017.02.021

[21] Osamusali SE, Groeneveld DC, Cheng SG. Two-phase flow regimes and onset of flow stratification in horizontal 37-rod bundles. Heat Technology. 1992; **10**:44-74. Available from: https://www.nrc.gov/docs/ML0425/ML042510526.pdf [Accessed: December 27, 2020]

[22] Yang J. An experimental study of flow stratification threshold in horizontal flow channels containing fuel bundles. In: 15th Int. Top. Meet. Nucl. React. Therm. NURETH-15, Pisa. 2012. Available from: https://inis.iaea.org/search/search.aspx?orig_q=RN:49101494 [Accessed: December 28, 2020]

[23] White FM. Fluid Mechanics. McGraw Hill; 2011. Available from: https://books.google.ca/books?id=egk8SQAACAAJ

[24] Pioro IL, Duffey RB. Heat Transfer and Hydraulic Resistance at Supercritical Pressures in Power Engineering Applications. ASME Press; 2007. Available from: https://books.google.ca/books?id=-cQeAQAAIAAJ

[25] International Atomic Energy Agency. Thermohydraulic Relationships for Advanced Water Cooled Reactors, IAEA-TECDOC-1203. Vienna: IAEA; 2001

[26] D'Auria F, editor. Thermal-Hydraulics of Water Cooled Nuclear Reactors. Elsevier; 2017. DOI: 10.1016/C2015-0-00235-0

[27] Martinelli RC, Nelson DB. Prediction of pressure drop during forced-circulation boiling of water. Transactions of the American Society of Mechanical Engineers. 1948;**70**:695-702

[28] Lockhart RW, Martinelli RC. Proposed correlation of data for isothermal two-phase two component flow in pipes. Chemical Engineering Progress. 1949;**45**:39-48

Chapter 3

Graphene-Based Functional Coatings for Pool Boiling Heat Transfer Enhancements

Aniket M. Rishi

Abstract

Pool boiling heat transfer has proven to be the most effective ways to dissipate the large amount of heat fluxes and achieve the efficient cooling in many industrial applications including high-power electronics cooling, data center cooling, heat exchangers, batteries, refrigeration, and air conditioning. With the aggressive net-zero carbon footprint goals set up by the numerous industries across the globe, the need for development of innovative two-phase cooling solutions is of utmost importance. Graphene, being the highest thermal conductivity material, has been implemented in numerous studies for improving both the critical heat flux (maximum possible heat removed before thermal runaway of the heater surface) and a heat transfer coefficient (determines how efficiently the heat is removed) in pool boiling heat transfer. Initially, this chapter introduces various graphene-based nanomaterials and basics related to structure and characterization of graphene. Later, the highlights of some of the notable research work related to the graphene-based coatings for pool boiling enhancements are discussed. The responsible mechanism for such higher performance is summarized. Concluding remarks and industrial applicability of these techniques are also discussed in this section.

Keywords: graphene-based coatings, pool boiling, efficient heat removal, two-phase cooling, advanced functional coatings, graphene composites

1. Introduction

Boiling heat transfer performance in any system primarily depends on the thermal conductivity of the heater surface as the heat is transferred from the heater surface to the fluid in contact. Higher thermal conductivity is desirable in achieving better heat removal efficiency. Various metals such as copper, gold, silver, and aluminum are commonly used in the industries as the heater surface materials due to their high thermal conductivity. After the discovery of world's thinnest graphene material by Geim and Novoselov in 2004 [1], numerous research studies have focused on implementing the graphene on heater surface to increase the boiling heat transfer performance. Graphene is a *2D* form of graphite and consists of a single layer of carbon atoms that are bonded together in a hexagonal lattice structure. Owing to its atomic layer thickness, a single layer of graphene possesses extremely high thermal conductivity, in the range of 4000–5300 W/m K. This chapter focuses on recent advancements in pool boiling heat transfer technique using graphene-based surfaces.

1.1 Graphene and its properties

1.1.1 Basics of sp^2-hybridization

The extremely high in-plane thermal conductivity possessed by graphene is primarily due to the sp^2-hybridized carbon atoms. In contrast, out-of-plane thermal conductivity is low because weak van der Waals interactions link the adjacent graphene planes within the multi-layer graphene.

A carbon atom consists of six electrons, and as per the energy states, two electrons are in 1 s state, and remaining four electrons occupy 2 s and 2p orbitals. In case of hexagonal structure of graphene as indicated in **Figure 1**, alternate carbon atoms are bonded via double bonds. Due to similar energies of 2 s and 2p orbitals, the electrons in these two orbitals arrange themselves such a way that one electron from 2 s orbital shifts in p orbit and contributes to forming three sp^2-hybrid orbitals, while a remaining one electron in p orbital forms a pi-bond with the neighboring carbon atom. Three sp^2 hybridized carbon atoms are bonded via a strong covalent sigma bond to other carbon atoms. **Figure 1b** indicates the sp^2-hybridization mechanism and the formation of 1 pi and three sigma bonds between two carbon atoms [2, 3].

1.2 Graphene-based nanomaterials

Although a single layer of graphene has the highest thermal conductivity, manufacturing techniques to produce a single layer graphene are expensive and time-consuming. This has led the researchers to develop various derivatives of graphene that are easier to manufacture and are comparatively cheaper. Some of these

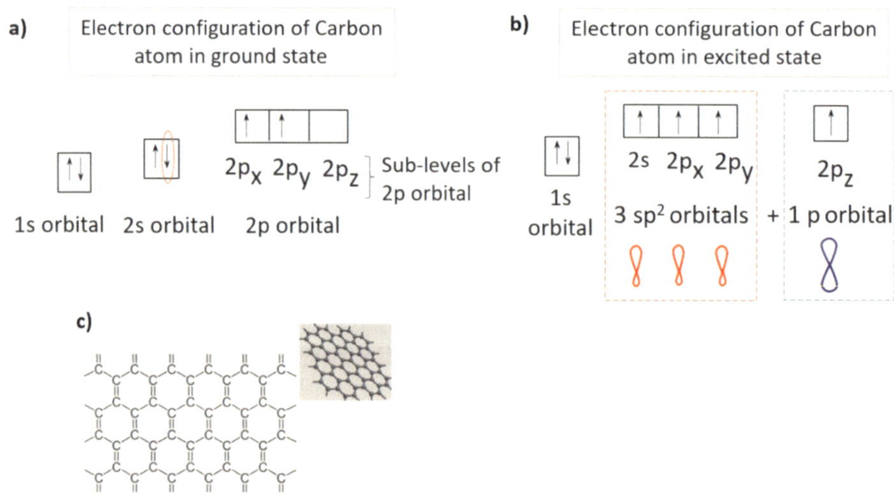

Figure 1.
(a) Distribution of electrons in a carbon atom in a ground state, (b) sp^2-hybridization of carbon atoms, and (c) structure of a single layer of carbon atoms forming a hexagonal lattice structure of graphene with sp^2 hybridized carbon atoms.

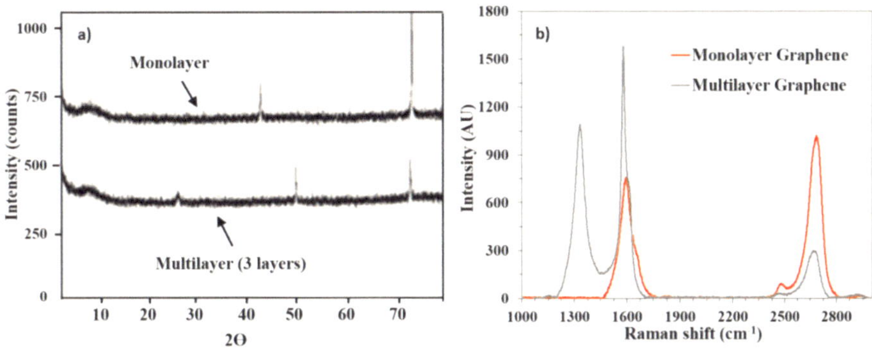

Figure 2.
(a) Typical X-ray diffraction peaks for graphene, (b) typical Raman spectra of a monolayer and multilayer graphene.

alternatives are graphene oxide (GO), reduced graphene oxide (rGO), and graphene nano-platelets (GNP). These forms of graphene, however, have reduced thermal properties due to additional functional groups attached to the carbon atoms and a loss of pure carbon structure. For example, GO has an additional oxygen (-O), carbonyl (=O), hydroxyl (-OH), and carboxyl (-COOH) groups attached to the graphene's hexagonal lattice of carbon atoms. Due to the presence of these groups, the thermal conductivity of GO is less than the pristine graphene, while the wettability of GO is higher. Graphene oxide can further be reduced to obtain a reduced graphene oxide (rGO) by eliminating the oxygen-based groups in GO. Defects in graphene structure also occur because of production methods that correspond to the breaking of the symmetry of honeycomb lattice carbon structure. Some of the defects include edge defect, grain boundaries defect, and defects associated with the change of hybridization of carbon from sp^2 into sp^3. The amount and nature of defects strongly depend on the production method and can have a large influence on the properties of graphene.

Another derivative of graphene known as GNP comprises a few layers of graphene tightly packed together in a hexagonal lattice structure. And despite their multilayer structure, GNP can yield thermal conductivity in the range of 2500–3500 W/m K, that is, comparable to a single layer of graphene. In addition, GNP have higher wettability than pristine graphene due to the presence of oxygen-based chains attached to the carbon atoms.

Numerous research studies have shown that as compared to the plain surfaces, increased wettability or hydrophilic surfaces/coatings yield higher pool boiling performance compared to both plain and hydrophobic surfaces. Thus, having a hydrophilic nature for the graphene-based derivatives provide an additional advantage in case of boiling heat transfer. This will further be discussed in detail later in the chapter.

1.3 Characterization techniques of graphene

Due to its atomic layer thickness, in pool boiling heat transfer applications, graphene must be deposited on the heater surfaces through various deposition techniques. Determination of the quality of deposited graphene layers and quantification of the number of deposited layers is very important for establishing the enhancement mechanism and validation of the deposition technique used. Poor quality of deposited graphene can also substantially affect its thermal and mechanical properties. And with increased number of deposited layers, graphene structure tends to become like a *3D* graphite structure that has lower thermal properties. Thus, it is important to understand the properties of the deposited graphene in any work that relies on properties of graphene for its superior performance. Most widely implemented, effective characterization techniques for graphene are discussed in this sub-section. These techniques are also of the utmost importance in determining the underlying enhancement mechanisms in pool boiling performance.

1.3.1 Raman spectroscopy

Raman spectroscopy is a light scattering technique in which molecules scatter incident light from a high-energy laser light source. Most of the scattered light have same frequency as the incident light; however, some fraction of light scatters at different frequency depending on the chemical structure such as benzene ring structure and bonds such as C=C, C—O, and C—H. Each peak in the Raman spectra corresponds to vibration of a specific molecular bond. The wavelength of the Raman scattered light depends on the wavelength of the incident light, and thus, Raman scatter wavelength number becomes impractical for the comparison. Thus, Raman scatter position is converted to Raman shift which indicates the Raman shift away from excitation wavelength.

Graphene-based derivatives typically show *D*, *G*, and *2D* peaks indicating the variation in peak intensity based on the quality and number of layers of deposited graphene. The distinct graphene peaks, *G* at $\sim$1580 cm^{-1} and *D* at $\sim$1340 cm^{-1}, correlate to the in-plane vibration of sp^2-hybridized carbon atoms and degree of disorder of sp^3-hybridized carbon structure, respectively. Monolayer graphene is generally defect free and thus does not show *D* peak (as shown in **Figure 2b**). *2D* peak is the second order *D*-peak that is observed at $\sim$2660 cm^{-1} and is a connotation of *D*-peak. Depending on the color, intensity, and type of wavelength of the laser, a small Raman shift can be observed on the x-axis. The ratios of *G* and *2D* peak intensities (I_G/I_{2D}) from Raman spectroscopy plot are used to find the number of deposited graphene layers. While the ratio of *D* and *G* peak intensities (I_D/I_G) represent the oxidation degree and defects on graphene sheets. It also represents the sp^3/sp^2 carbon ratio. Generally, (I_D/I_G) ratio of less than 1 indicates the good quality and less defects on the graphene structure and thus has lesser impact on its thermal and mechanical properties.

1.3.2 X-ray diffraction (XRD)

X-ray diffraction is a technique used for determining the atomic and molecular structure of a crystal in which crystalline atoms cause a beam of X-rays to diffract in specific directions. When X-rays are incident on the sample, the incident beam gets separated into transmitted beam and diffracted beam. The diffraction pattern is recorded in terms of 2θ angle that indicates the crystalline phase of the material. The crystalline phases of graphene are typically investigated using an X-ray diffractometer (XRD) with Cu Kα radiation; wavelength 1.5418 Å. The spectra are recorded for 2θ ranges between 5° and 75° at a rate of 3°/min rate. The step size is 0.02° with an X-ray power of 40 kV and 35 mA. This range captures peaks from carbon and the underlying copper substrate in case of graphene deposition on copper. The location of characteristic peaks determines the presence of elements on the surface.

X-ray diffraction 2θ reflection peaks between 6° and 10° correspond to graphene. Peaks are typically either broader or sharp between 6° and 10°. **Figure 2a** shows the comparison of the XRD plot for monolayer and multilayer graphene coatings deposited on copper substrate. The peak intensity is on the y-axis, and 2θ reflections peaks are on the x-axis. The peaks confirm the presence of graphene along with copper. With the presence of large amount of carbon and more disordered structure, additional G peak at 20° can also be prominently observed for XRD plots.

2. Applications of graphene in pool boiling heat transfer

As discussed previously, owing to extremely advantageous thermal properties of graphene and graphene-based derivatives, numerous research studies have focused on using the graphene for enhancing the pool boiling heat transfer performance. Less complex, less hazardous, and less time-consuming deposition techniques are desired for environmental, economic, and industrial purposes. This section is divided into two main sub-topics: (1) Nanoscale graphene-based coatings and (2) Microscale graphene-based composite coatings. Various deposition techniques of graphene on the heater substrates are discussed along with their corresponding pool boiling heat transfer efficacies in the subsequent sections. The enhancement mechanisms responsible for such heat transfer improvements are also discussed. Amongst numerous research work available, the studies have been selected based on fulfillment of either one or all the following criteria: higher pool boiling performance, potential for applicability in industrial applications, and development of unique/innovative approach in creating graphene-based functional coatings.

2.1 Nanoscale graphene-based coatings

Nanoscale coatings, as the name suggests, have thickness in the range of nanometer and mainly consist of only graphene as the depositing material. Some of the widely implemented techniques to develop nanoscale graphene-based coatings are discussed here with the focus on their applications in pool boiling heat transfer enhancements.

2.1.1 Chemical vapor deposition (CVD)

Even though a variety of techniques exist focusing on production of high-quality single layer graphene, the challenges related to the transfer of graphene on the heater substrate after production still remain a significant obstacle for its industrial adoption. Chemical vapor deposition technique is a step toward the application of graphene in industries as a graphene layer can directly be formed on the heater substrate.

Generally, to generate the graphene film, an atmospheric pressure CVD (APCVD) technique is used. The polished copper substrate is loaded in the tubular furnace and vacuumed. The temperature is then ramped to 1333 K at the rate of 303 K/min. With a mixed gas of H_2/Ar (20/80 standard cubic centimeters per minute, (sccm)) at a constant pressure of 72 kPa. To flatten and to reduce the copper surface, an additional step of surface annealing is performed under a mixed gas of H_2/Ar (30/1000 sccm) at 101,325 kPa for 30 minutes. During the growth of graphene, the atmosphere is switched to CH_4/H_2/Ar (0.5/30/1000 sccm) at 101,325 kPa for 7 minutes. Followed by this, it is cooled down to room temperature under H_2/Ar (30/1000 sccm) flow to complete the process of deposition and to avoid the oxidation of the deposited film [4]. Bulk copper substrate in the CVD introduces the roughness effects due to thermal deformation. This provides additional morphological features and is beneficial for boiling heat transfer. A monolayer and multilayer graphene coated surfaces are developed using this APCVD technique. Raman spectroscopy confirms the quantification of the deposited layers. Compared to the plain copper surface (critical heat flux (CHF) = 1280 kW/m^2) using distilled water as a working fluid, both monolayer and a multilayer (three layers) graphene coatings yielded higher pool boiling performance with CHF of 1490 kW/m^2 and 1570 kW/m^2, respectively. The improved performance is attributed to the altered liquid wettability along with the wrinkle induced roughness of the underlying copper that provides additional nucleation during the boiling [5].

Another study is performed by implementing APCVD to create graphene-based coatings on larger heater sizes (32 × 32 mm as against 10 × 10 mm) and HFE-7000 refrigerant as a working fluid instead of distilled water. A similar gas composition of $CH_4/H_2/Ar$ are used to develop the coatings. Additional fluorinated-graphene coating is prepared by gas fluorination of CVD-grown graphene in a Teflon container with XeF_2 powder as a precursor (loading mass: 0.2 g per 0.4 L) followed by baking in the oven at 373 K for 12 hours. Compared to the plain copper surface, 2 times higher bubble density and 2.43 times higher bubble density is observed on CVD grown graphene and fluorinated-graphene coatings. Increased hydrophobicity of the coatings is mainly responsible for the increased bubble activity. This resulted in 80 and 20% higher heat transfer coefficient (HTC) for CVD grown graphene and fluorinated-graphene coatings as compared to the plain copper surface (HTC = 5.3 kW/m^2 K). However, due to the increased bubble density, the CHF for CVD grown fluorinated-graphene coating is 370 kW/m^2, which is lower than the plain copper surface (CHF = 380 kW/m^2), while CVD graphene shows the highest CHF of 428 kW/m^2 [6].

Various other similar studies have been performed using APCVD methods to create graphene-based coatings for the improvement of pool boiling performance. One of the studies have created hybrid coated surfaces using graphene and carbon nanotubes for improvements in pool boiling. Similar improvements in pool boiling performance have been observed with these studies. However, longevity studies from the industrial applications perspective have been performed on CVD graphene coating methods for pool boiling heat transfer.

2.1.2 Nanofluids

To create graphene-based nanoscale coatings, researchers have utilized graphene (G)/graphene oxide (GO)/reduced graphene oxide (rGO) based nanofluids. These coatings are formed when the nanofluids are boiled on the substrate surfaces for a specific amount of time. This nanoscale coated surfaces are then utilized as the substrates on which pool boiling is performed. Typically, these developed coatings are self-assembled and have varied morphological features that assist in improving the pool boiling performance.

In one of the studies, a new concept of nucleation patterning surface along with rGO-coated pillars structures are developed. The micropillar structures are developed using UV lithography technique in which a pattern mask is created with the size of the pillars. Deep reactive ion etching is then performed to generate the micropillars. The diameter of 4 μm, a pitch of 20 μm, and a height of 20 μm are the dimensions of the micropillar structures. Additionally, to improve the nucleation performance during boiling, a micropillar free region of 200 × 200 μm was also kept. The prepared surfaces were then coated by boiling rGO using 0.0005 wt.% rGO nanofluid solution and stopped just before reaching the CHF. At this stage, the self-assembled structures of rGO on micropillars are obtained. Two different surfaces with pitch of 1 and 1.5 mm are developed to minimize the bubble coalescence. These coatings are implemented for pool boiling studies using de-ionized (DI) water as a working fluid. Compared to the CHF on a plain silicon chip (890 kW/m^2), 1.5 pitch rGO-coated micropillar structure reached the CHF of 2700 kW/m^2 along with HTC of 89.7 kW/m^2 K (plain surface HTC = 20 kW/m^2 K). Plain silicon surface with rGO coating is also tested for the pool boiling which yields CHF of 1340 kW/m^2 and the HTC of 80 kW/m^2 K. The rGO-coated micropillar-free cavity on the surfaces facilitated bubble nucleation by providing cavities on the basis of bubble departure diameter which delayed the horizontal coalescence of the bubbles and increased the overall pool boiling performance. The micropillars provided the liquid paths to the nucleation cavities and rGO layer on top of the micropillars ensured the continuous capillary inflow [7, 8].

Another study is performed with varying the concentration of graphene in graphene-based nanofluid. The pool boiling is conducted on the copper substrates, and a maximum of 46% reduction in wall superheat is observed for 0.2% graphene nanofluid. Along with this, a maximum HTC improvement of 48.6% is observed as compared to the plain copper surface with DI water. This enhancement is attributed to the possible deposition of graphene at higher heat fluxes on the substrates and producing of additional nucleation sites that assist in boiling [9].

In one of the studies with the rGO, the effect of base-graphene layer, self-assembled foam like graphene layer and a thick-graphene layer are compared for their pool boiling performance on SiO_2 surfaces. These nanofluid-based coatings are generated by boiling the nanofluids on the boiling surface, followed by the actual boiling test. Different types of layer formation on the substrate depends mainly on applied heat flux and the concentration of the rGO colloid. The CHF is increased with reducing the concentration of rGO colloidal solution. A maximum CHF of 1600 kW/m^2 is achieved for 0.0001 wt.% of rGO. It is observed that during water boiling, the thick-graphene layer coating was completely detached and the portions of base-graphene layer and self-assembled foam like graphene layer remained and assisted in increasing heat spreading actions. Additionally, it is observed that even after reaching CHF, the substrate surface temperature increased very slowly, and the heat flux is maintained. It is hypothesized that this is due to the heat spreading action by base-graphene layer and porosity introduced due to the deposition [10].

The effect of rGO nanofluids is also studied on the copper flat plate heater. Different rGO wt.% of 0.2, 0.6, and 0.8 are considered, and since the thickness of the plate heater is 0.05 mm, initial step of developing a coating by boiling a nanofluid is not performed. The tests are performed till the CHF with different rGO concentrations and are compared with the plain DI water. The highest CHF of 945 kW/m^2 is attained for 0.2 wt. %. However, the HTC for 0.8 wt.% rGO is the highest due to increased nanoparticles in the DI water and their deposition to form a porous layer on the boiling surface that acts as secondary cavities. For the heat flux of 500 kW/m^2, 2.3 times higher HTC is obtained for 0.8 wt.% rGO than the boiling with DI water on copper plate heater [11]. Uncertainties of measured and calculated parameters are considered while reporting all the CHF and HTC values that have been mentioned throughout the chapter [12–14].

2.1.3 Concluding remarks

Many of similar studies have been performed to develop the nanoscale graphene-based coatings either by nanofluids or by chemical vapor deposition techniques [15, 16]. The enhancement mechanisms responsible for these improvements can be categorized into either one of the following or the combination of the following: Nanoscale coatings with graphene alter the wettability of surfaces, enhance the thermal conductivity, introduce additional surface roughness features, and increase the nucleation activity during boiling. Even though many innovative approaches have been introduced over the years, not many studies have focused on longevity of these surfaces for real-world industrial applications. Also, compared to microscale composite coatings, nanoscale coatings do not yield higher pool boiling performance and lack the ability to sustain continuous vigorous boiling over the longer periods. Considering the continuous usage and complications in the industrial applications, these nanoscale enhancement techniques appear to have limited scope.

2.2 Microscale graphene-based composite coatings

Pool boiling performance enhancements using deposition of composite coatings have been reported extensively over the years. Composite coatings represent the deposition of a material using more than one metal and/or non-metal on the heater substrate by means of various deposition techniques. And these composite coatings typically attain higher critical heat fluxes at lower wall superheat temperatures due to the formation of porosity which results in increased surface area and bubble nucleation sites available for boiling.

To be effective during the pool boiling, the bond strength of the porous coatings with the base substrate must be strong enough to sustain the vigorous boiling. Otherwise, coatings can peel off from the heater surface substrate and can result in thermal runaway. This bond strength and adhesion is observed higher in composite coatings than pure coatings since metal-to-metal diffusion bonds have higher mechanical properties. The process and mechanism of vapor bubble generation in the porous surfaces is as follows: When the heat is supplied to the substrate/heater surface, above a certain temperature, the nucleus of a bubble grows in the cavity. The cavities may be formed inside the porous matrix or may appear on the surface of the coating. When this bubble nucleates and grows, it carries heat with itself. As the bubble departs from a cavity after reaching its critical radius, the liquid in the vicinity of the void rushes into the cavity, ensuring a continuous supply of liquid for evaporation. Steady vapor formation takes place on the porous media, and the nucleation takes place within the porous matrix via the re-entrant cavities that are not susceptible to flooding by liquid. Higher the nucleation frequency, higher is the heat dissipated from the surface. And smaller the vapor bubble diameter, more is the bubble generation and thus higher is the heat removal. Further, the agitation caused by the bubble activity may increase the heat transfer rate between the surface and the liquid.

Compared to nanoscale coatings, microscale coatings have many advantages from industrial applicability perspective such as higher cohesive and adhesive bond strength, higher pool boiling performance, and ability to sustain repetitive boiling over a prolonged time. Electrodeposition, dip coating, and sintering are some of the simple and widely used deposition techniques to develop the microscale coatings not only for pool boiling enhancement applications but also for a variety of industrial applications, including nuclear power plants, water desalination, vapor chambers, electrodes of lithium-ion batteries, and condensation heat transfer. Detailed studies related to these coatings are discussed in this section.

2.2.1 Electrodeposition technique

Electrochemical deposition is the process of coating solids on the conductive base materials to modify the surface properties. It consists of an electrolyte solution containing positive and negative ions usually prepared from metal salts and the two working electrodes that can be of either conducting or semiconducting nature known as the cathode (on which the coating is desired) and an anode. The resultant electric current (rate of the motion of the electric charge) between the two electrodes under an external voltage is due to the migration and diffusion of the charged species. And since it is difficult to diffuse graphene and graphene-based nanomaterials in the metal base substrate, typically in pool boiling applications, a metal salt is also vital to deposit a composite mixture of the metal and graphene. This provides a very high bond strength with the substrate along with deposition of graphene that can both sustain the boiling for longer duration and maintain the higher heat transfer performance.

The principle behind electrodeposition is to use an electric current to strip the cations from a sacrificial material (anode) in a solution and coat that material in the form of a thin film onto a substrate that is conductive (cathode). The electrodes are positioned parallelly in the electrolyte solution containing both positively charged ions called cations and negatively charged ions called anions. When an external electric field is applied, the cations depart toward the cathode and get deposited as metal. According to which, the process follows Faraday's law, and the amount of the deposited metal on the electrode is proportional to the applied current to the electrochemical cell.

$$Q = \frac{n * d * A * h * F}{M} \tag{1}$$

Where Q is the charge (C), n is number of electrons, d is the distance between cathode and anode (m), A is the coated surface area (m^2), h is the thickness of the deposition (m), F is Faraday's constant, and M is the atomic weight.

In this work, the duration of applied current is also considered.

$$Q = I * t \tag{2}$$

Where Q is the charge applied (C), I is the amount of current supplied (mA), and t is the duration for which the current has been provided (seconds).

For the application of pool boiling heat transfer, a coating with defects and pores is desired due to its tremendous advantages that include pores acting as nucleation sites, increased wicking and wetting properties, and non-uniformity of the porous structures that serve as a perfect platform for very intense bubble formation and heat dissipation. Addition of graphene to such porous structures can further amplify the performance due to additional increased in-plane thermal conductivity of the coatings and formation of unique morphological structures of the coating.

A multi-step electrodeposition technique includes multiple steps of varying current and time durations while performing the electrodeposition process. Each step has specific values of controlling parameters. The electrodeposition technique can be implemented on the substrate of any shape, size, and material. And its applications are not only limited to boiling heat transfer substrates, but also include any electroplating and coating-based applications. Some of the target applications include abrasion and wear resistance protection, corrosion protection, decorative coatings, prolonged life of coating and surface, durability, to maintain the esthetics, integrated electronics, solar reactors, fabrications, and others.

2.2.1.1 Dynamic template-assisted electrodeposition technique

Template-assisted electrodeposition permits more size and shape-controlled deposition. The templates could either be dynamic, restrictive, or self-organized. In the case of an aqueous solution, where electrolysis of water takes place in one of the electrochemical reactions, the evolved hydrogen serves as the dynamic template that results in porous surface coatings. The electrochemical reaction takes place when direct current is supplied through the electrochemical cell. First, at a higher current density supply, hydrogen bubbles are formed on the cathode (substrate). Typically, the electrolysis of water in the electrolyte creates hydrogen gas. If the evolution of bubbles is continued, the copper ions start to grow within the interstitial spaces between hydrogen bubbles. The resultant hydrogen bubbles behave as dynamic templates around which copper particles deposit and grow. When the higher current density supply is stopped, the hydrogen gas bubbles collapse, leaving behind the porous open network of copper. The size of the pores is determined by the bubble behavior, and the morphology of the metal film is determined by the nucleation and growth mechanism of the metal on the substrate. Thus, two simultaneous reactions occur at the cathode, deposition of copper ions and the evolution of hydrogen bubbles.

The hydrogen evolution reaction rate depends on the applied current density which in turn is dependent on the depositing metal-hydrogen chemisorption energy or the dissociation of hydrogen ions from the electrode surface that further combines with protons to form hydrogen gas. This is dependent on the current exchange density, which is defined as the rate of hydrogen evolution per surface area at the electrode. Metals that have weak interaction energy with hydrogen do not abet in the passage of sufficient electrons, whereas metals that interact strongly with hydrogen result in greater adherence to the surface and not getting released in the solution instantly.

The applied current density and duration control the surface morphology characteristics such as porosity, thickness, wickability, wettability, contact angle, hydrophilicity, and hydrophobicity of the electrodeposited coatings. Bubble behavior dictates the size of the pores. A higher bubble generation rate leads to shorter residence times that further control the coalescence of the bubbles. Reduced coalescence results in the smaller pore sizes of the deposited materials. Effectively, higher current density results in the production of the higher amount of hydrogen ions that result in the formation of hydrogen bubbles which provide the anatomy for the porous network and subsequent higher amounts of metal ions produced in this step get deposited around these hydrogen bubbles.

According to Faraday's law of electrolysis [17], the amount of deposited material is proportional to the duration and time of the deposition. The deposition is usually accomplished using two methods: (i) supplying the constant current during the deposition process and (ii) holding the constant potential during the electrodeposition. These methods are called as galvanostatic and potentiostatic modes of deposition, respectively. In the current study, all the depositions are performed using the galvanostatic method, i.e., the constant current is supplied for the fixed duration, and the voltage is varied accordingly. By controlling the current density and time required for the deposition, the morphological structure and porosity of the coating can be controlled using the electrodeposition technique.

2.2.1.1.1 G/GO-Cu composite coatings

Copper being highly thermally conductive metal, deposition of G/GO-Cu composites is performed using the copper block as anode and a plain copper substrate as the cathode with addition of G/GO colloidal solution by % by volume. Both the electrodes are held parallel by placing in a Poly-Tetra-Fluoro-Ethylene (PTFE) holder, and the entire assembly is placed in the electrolytic bath consisting of 5.85 gm of 0.8 Molar concentration of $CuSO_4$, 3.14 mL of 1.5 Molar concentrated H_2SO_4, 40 mL distilled water, and 0.5, 1, 1.5, and 2.5% vol./vol. G/GO solution. The working area on both the electrodes is delineated with Kapton® tape (**Figure 3**).

To achieve a microporous coating on the heater surface, template assisted electrodeposition technique is adopted which includes supply of higher current density of 400 mA/cm^2 for 15 seconds that produce hydrogen gas bubbles as a result of electrolysis of water and the deposition occurs around these evolved bubbles. Lower current density supply step of 40 mA/cm^2 for 2500 seconds after high current density step

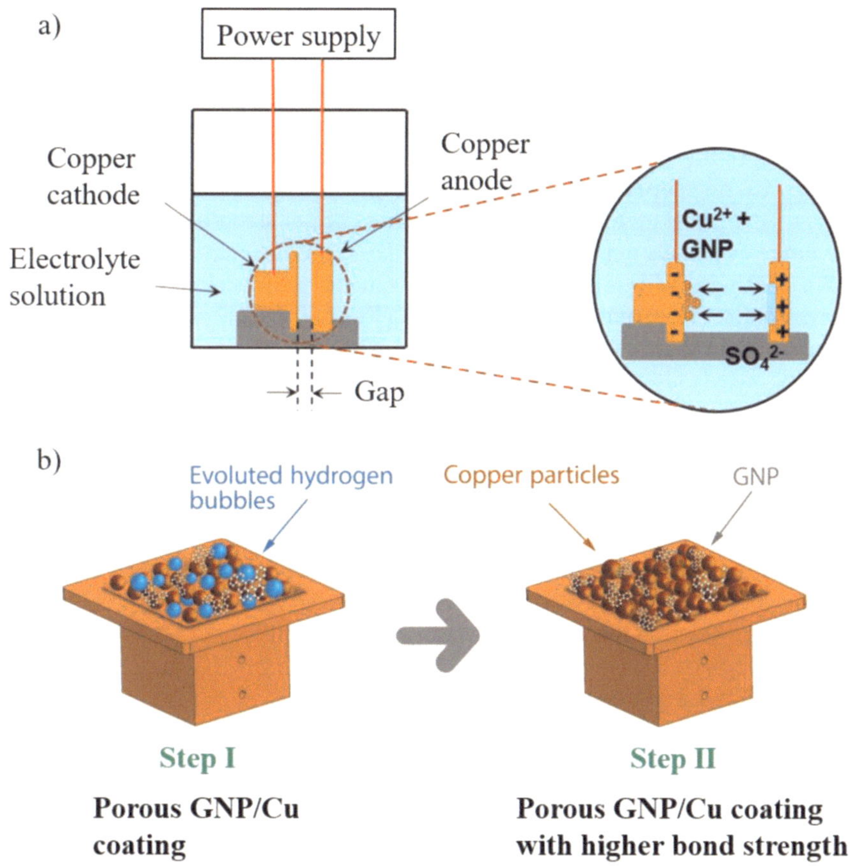

Figure 3.
(a) Typical setup for deposition using electrodeposition technique, (b) two-step electrodeposition technique schematic with copper and GNP.

deposits a small quantity of G/GO-Cu without evolution of hydrogen bubbles and strengthens the adhesion of the coating on the substrate. The formation and collapse of hydrogen bubbles during two-step electrodeposition technique ultimately yields highly microporous coating. Compared to the plain copper surface (CHF = 1280 kW/m^2) using distilled water as a working fluid, all Cu-G/GO vol. % composite coatings yielded higher pool boiling performance, with 2.5% G/GO coating giving maximum CHF of 2200 kW/m^2 and the HTC of 155 kW/m^2 K (as compared to 55 kW/m^2 K for plain copper surface). The enhancement mechanism for achieving higher performance can be condensed to the combination of following multiple factors—increase in wetting and wicking properties, wicking through dendrite type copper structures and bubble nucleation on underlying deposited G/GO sheets, increased nucleation sites that become activated under suitable substrate temperature conditions and contribute toward decreasing the wall superheat, and enhanced evaporation through microlayer partitioning mechanisms that increase bubble growth rates and frequency [18].

2.2.1.1.2 GNP-Cu composite coatings

A similar electrodeposition technique as that of G/GO-Cu composite coatings is also implemented to develop GNP-Cu composite coatings considering higher thermal properties of GNP than G/GO. GNP powder is commercially available and is added by varying the wt. %, 0.25, 0.5, 1, 2, 2.5% in the electrolyte solution. All the electrochemical parameters and electrolyte solution composition are kept constant to compare the difference between G/GO and GNP. The electrodeposited coatings are then tested for pool boiling heat transfer with water as working fluid till the CHF condition.

Compared to the G/GO-Cu composite coatings, a very distinguished morphological features are developed for GNP-Cu composite coatings. Due to their multilayered structures, GNP sheets wrap and deposit around copper structures and the substrate. A wide range of porous network is developed for GNP-Cu coatings as against the G/GO-Cu coatings. Additionally, all the GNP-Cu electrodeposited coatings are superhydrophilic (0° contact angle) in nature (**Figure 4**c,d). For 2% GNP-Cu coating, a maximum wicking rate is attained, with 2 μL water droplet wicking within 12 ms. The wicking rate (unit - m/s) that is calculated by normalizing the wicked volume over the droplet contact area is 0.145 m/s, maximum for 2% GNP-Cu coating, while lowest rate of 0.018 m/s is obtained for 0.5% GNP-Cu coating. The pool boiling studies show an overall increase in both CHF and HTC for all GNP concentrations. Pool boiling and HTC plots are presented in **Figure 4**a,b. CHF of 2670, 2400, 2860, and 2750 kW/m^2 is achieved for 0.5, 1, 2, and 2.5% GNP/Cu coatings, while 0% GNP-Cu coating (only copper coating) achieved a CHF of 1560 kW/m^2. Heat transfer coefficients of 142, 194, 204, and 150 kW/m^2 K are rendered for 0.5, 1, 2, and 2.5% GNP-Cu coatings, while 0% GNP-Cu coating yielded an HTC of 60 kW/m^2 K. 2% GNP-Cu coating yielded maximum of ~130% increment in CHF and 290% increment in HTC as compared to the plain copper surface.

A unique shift in pool boiling curve toward the left (or lower wall superheat) with increase in heat flux is observed for GNP-Cu coatings as observed in **Figure 4a**. This shift is primarily due to the activation of additional nucleation sites for boiling at different wall superheat temperatures, and this phenomenon is termed as "boiling inversion". The underlying mechanism for boiling inversion in porous surfaces is attributed to the presence of hierarchical pores that develop supplementary nucleation cavities at higher heat flux, and thermally induced gradients along the pores also dominate owing to varying thermal conductivity of the material.

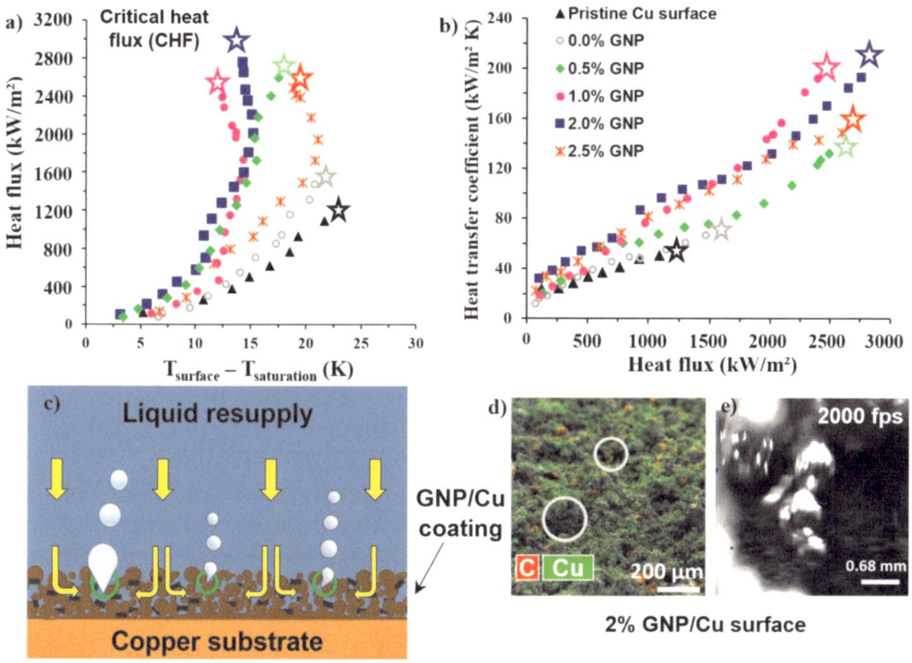

Figure 4.
(a) Pool boiling curve, (b) heat transfer coefficients for GNP-Cu coatings using distilled water as a working fluid at an atmospheric pressure (size of the boiling surface 10 mm × 10 mm), (c) schematic of enhancement mechanism for GNP-Cu coatings, (d) porosity on the GNP-Cu electrodeposited coating, and e) stream of departing bubbles from nucleation cavity at 100 kW/m² for 2% GNP-Cu surface.

The presence of hierarchical pores also improves the CHF and HTC due to added nucleation sites for generating additional vapor bubbles that improve the boiling efficacy. Superhydrophilic coatings further promote the nucleation and microlayer evaporation during the boiling. Visualization studies indicate very low bubble departure diameter of 0.68 mm for 2% GNP-Cu coating along with lower departure time of 5 ms as shown in **Figure 4e**. The combination of all these factors along with higher thermal conductivity of GNP forms the enhancement mechanism and yields the highest pool boiling performance ever recorded on plain copper surface with GNP [19].

2.2.2 Sintering technique

Sintering is the process of compacting a powdered material and forming a solid or porous coherent mass by heat or pressure without melting it to the point of liquefaction. It is a heat treatment process that is generally used to increase the strength and structural integrity of the material. The produced coating provides a higher surface area-to-volume ratio compared to its bulk counterpart. Sintering strengthens the particle contacts by means of the thermal mass transport process and provides the change in porosity and pore geometry. Sintering technique has many advantages over other coating processes—controlled deposition for tunable coating thickness; the ability to coat substrates of varying shapes and thickness; and cost-effectiveness.

The sintering process is an irreversible event that happens in different stages. The sintering temperature is high enough to promote neck formation at the point of contact between the adjacent metal particles. The process initiates with loose powders with a specific packing density if no compression is involved, as shown in **Figure 5a**. Initially, necks between the contacting particles grow to the point where the neck size is less than one-third of the particle size. During the next transitional stage, with the continuation of necking amongst the contacting particles, tubular pores start forming and connect to the external surface. Finally, the necking state is achieved to a point where only a small porosity is present in the material. The grain boundaries are developed at the neck regions. The porosity is an inherent property during a sintering process and can be altered by changing the sintering time. The schematic of the different stages is shown in **Figure 5**.

The porosity of the sintered coating primarily depends on the sintering temperature, the ratio of powder to sintering oil, and sintering time. Effectively, a higher temperature can distort the shape of the particles. And a lower sintering temperature can result in a coating that cannot develop enough bonding between the particles due to low temperature. This can reflect in poor bond strength and removal of the coatings. Similarly, if the sintering time is less, the bond strength of the deposited coating is inadequate and fails to sustain vigorous forces. And if the sintering time is more, it can reduce the porosity drastically. Thus, with the help of sintering parameters, the surface morphological characteristics such as porosity and thickness and the subsequent surface properties such as wettability and wickability can be controlled.

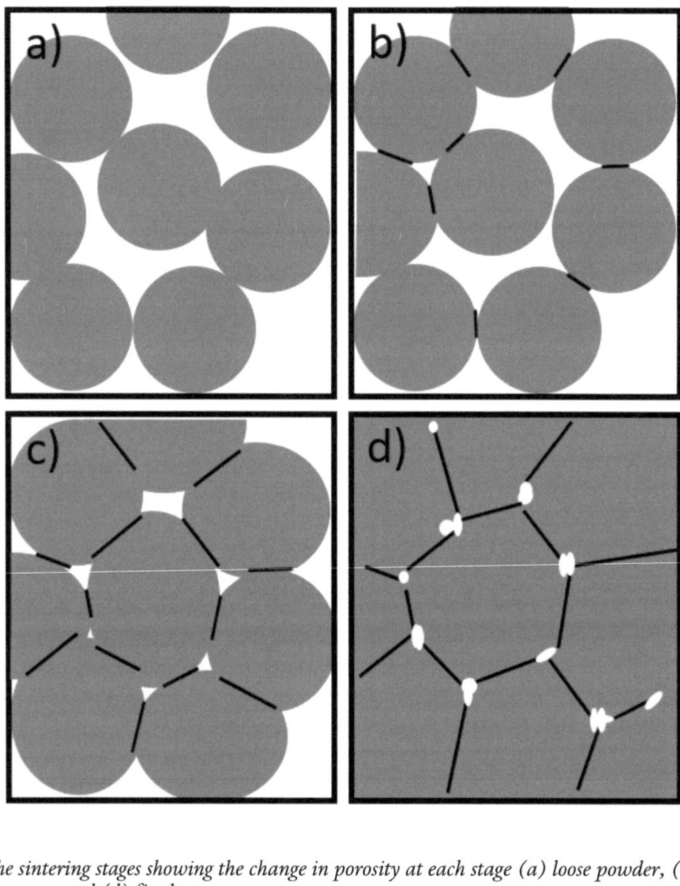

Figure 5.
Illustration of the sintering stages showing the change in porosity at each stage (a) loose powder, (b) initial stage, (c) intermediate stage, and (d) final stage.

2.2.2.1 G/GO-Cu composite coatings

Copper being highly thermally conductive metal, deposition of G/GO-Cu composites is performed using the copper block as anode and a plain copper substrate as the cathode with addition of G/GO colloidal solution by % by mass.

The produced coating provides higher surface area-to-volume ratio compared to its bulk counterpart. Formation of artificial nucleation cavities to promote vapor generation rate can be achieved using sintering technique. Here, the screen-printing paste is created by adding a commercially available screen-printing binder. The composite G/GO-Cu powder is mixed with the sintering oil with powder-to-oil ratio of 2:1. After screen-printing, the test surfaces are securely placed inside the sintering furnace with the inert helium atmosphere. During sintering, initially the sintering temperature is raised to 723 K for the duration of 30 min. to eliminate the binder from the coating and then is ramped up to 1073 K for a duration of 1 hour to develop a microporous sintered coating. The temperature is then ramped down to the room temperature via natural convection, and the sintered surfaces are removed from the furnace once the room temperature is attained. Results indicate the improvement in HTC for 0.2, 0.4, 0.6, and 1% G/GO % by mass than the plain copper surface. Critical heat fluxes obtained are similar to plain copper surface, with maximum CHF of 1420 kW/m^2 and the HTC of 194 kW/m^2 K attained for 1% G/GO-Cu coating. Small increments in CHF values are due to the poor wickability of the coatings and formation of hydrophobic coatings. Higher HTC is due to early nucleation activity on the surface and increased porosity [18].

Another study focused on the effect of foam thickness and graphene coating on pool boiling heat transfer of sintered porous surfaces. This study combined sintering technique for copper deposition and CVD process for monolayer graphene deposition on sintered copper deposits. Thinner foam thicknesses (0.5 and 1 mm) yielded better performance than 1.5 and 2 mm due to lower vapor resistance and efficient liquid supply. Heat transfer coefficients are observed 161% higher initially at lower heat fluxes, while this increment in HTC is reduced at higher heat fluxes [20].

2.2.2.2 GNP-Cu composite coatings

Traditionally implemented sintering techniques yield the coatings with uniform porous structure throughout the coating. However, the uniform spreading of graphene in the sintered coatings is not guaranteed and thus can affect the overall pool boiling performance of the coatings. In this work, to provide a homogeneous powdered mixture of copper and GNP, ball milling is performed prior to sintering. And this homogeneous composite powder is then used for sintering to achieve uniform spreading of powdered GNP. High-energy ball milling is selected due to its cost effectiveness in forming homogeneous powdered mixtures of composite materials and alloys. It is hypothesized that the ball milling enables draping of highly thermally conductive GNP around the copper particles and sintering with these GNP-draped-copper particles will result in microporous coatings with enhanced wetting and wicking properties, which will improve the pool boiling heat transfer performance.

During this ball milling process, a composite mixture is repeatedly cold welded, fractured, and re-welded to yield a homogenized powder. Before loading the composite particles mixture, to achieve a homogeneous mixing, GNP and copper particles are dispersed in the ethanol bath for 30 min. Along with stainless-steel balls, the entire solution is then transferred into the ball milling chamber, and the ethanol is used as a process control agent during the ball milling process. The ball-to-powder ratio of 40:1

is used to ensure the homogeneous distribution of GNP in the copper particles. The entire mixture is ball milled at 700 rpm for 1 hour. After every 15 minutes of ball milling, to avoid the overheating, the ball milling chamber is allowed to cool down for 1 hour. The resultant ball milled composite particles are then screen printed on the copper test surfaces using the sintering oil.

The collision of stainless-steel balls during the ball milling traps the GNP and copper particles (shown in **Figure 6**). The force of the impact flattens and plastically deforms the particles which lead to the work hardening and fracturing. Due to this, increment in surface-to-volume ratio of the particles is obtained. The repetitive ball-to-wall and ball-to-ball collisions during ball milling reduce particle size via fracturing and the cold welding, i.e., draping of GNP on copper particles. Annealing cycle of 1 hr. in between the ball milling relives the internal stresses and defects of GNP caused due to continuous collisions. In addition to particles size reduction, a uniform distribution of GNP around each copper particle facilitates, leading to increased wicking rates of the coatings [21].

Amongst different copper particles sizes and GNP concentrations, the 2% GNP with 20 μm Cu particles size performed the best yielding CHF of 2390 kW/m^2 at wall superheat of 8.4 K. This performance with ball milling followed by sintering is higher than both plain copper surface and only sintered Cu-GNP composite. Combination of the following mechanisms yielded this high performance: superhydrophilic coatings leading to increased microlayer evaporation and improved liquid supply to nucleation cavities, homogeneous mixture formation of GNP-Cu powder due to ball milling, increased thermal conductivity of the coatings resulting from the usage of GNP [22].

Similar study is performed using aluminum powder particles and developing Al@GNPs coatings using sintering with ball milled powder of aluminum and GNP. Aluminum substrate is used as a heater surface with R-134a refrigerant as a working fluid. With increase in coating thickness, pool boiling performance is improved, with the highest improvement in HTC of 143% observed for 125 μm thick coating than a plain aluminum surface. This enhancement is primarily due to increased nucleation sites, coating thickness and porosity [23].

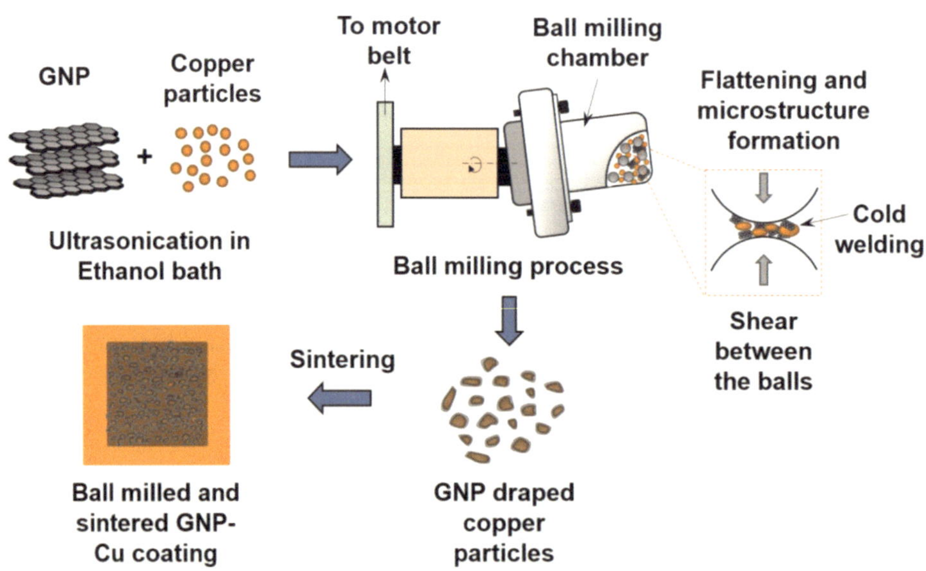

Figure 6.
Typical process of developing the ball milled powder and sintered coating using the ball milled composite powder.

2.2.2.3 GNP-Cu composite coatings with salt-templated sintering

Even though the performance of ball milled sintered coatings is higher than the plain copper substrate, such sintering techniques are limited by their control over the resultant morphological features, such as porosity and pore diameter that play a crucial role in determining the overall pool boiling efficacies of the coatings. This also limits the consequential surface properties such as wettability and wicking behavior of the coatings. In this study, the focus is provided on increasing the control of various surface properties such as porosity, wettability, and wickability. During sintering process of ball milled powder, sodium carbonate (Na_2CO_3) salt pellets are used here as the templates to achieve wide range of porous network. And post sintering, the sintered surfaces are rinsed with distilled water to dissolve and remove the traces of these salts. This non-uniform highly microporous structure is then used for pool boiling studies.

Pool boiling performance is evaluated after characterization of the surfaces. A dramatic increment in CHF is observed for the combined ball milled and salt templated sintered surfaces. A maximum CHF of 2890 kW/m^2 is attained for 20 μm Cu-3% GNP surface, which translates to ~131% enhancement in CHF than a plain copper surface. Besides, the wall superheat of just 2.2 K is achieved, representing ~2390% improvement in HTC than a plain copper surface. These are the highest CHF and HTC values reported in the pool boiling literature for graphene-based and porous coatings coated on a plain copper surface. For 0, 2, and 5% GNP coatings, CHF of 1550, 2690, and 2670 kW/m^2 are achieved, respectively. The maximum HTC of 1314 kW/m^2 K is obtained for 20 μm Cu-3% GNP coating, while HTC of 227, 399, and 431 kW/m^2 K are achieved for 20 μm Cu-0, 2, and 5% GNP coatings, respectively [22].

Similar to ball milled and sintered coatings, here the combination of increased thermal conductivity of coatings due to GNP-Cu ball milling and formation of superhydrophilic coatings with increased microlayer evaporation and improved liquid supply to nucleation cavities assisted in achieving the drastic enhancements in pool boiling performance. Apart from this, the key factors responsible for the improvement of both CHF and HTC are as follows: Formation of hierarchical microporous structure creates a wide range of porosity ranging from ~2 to ~200 μm. This wide range of developed pores serve various functions: Different range of pores activate at different wall superheat/heater surface temperature and initiate bubble nucleation activity. While the cavities not in the range of nucleation activity (primarily formed via the salt templating) act as liquid reservoirs and provide continuous liquid supply as soon as the vapor bubble departs. This provides massive advantage and thus has attained the highest pool boiling performance for graphene-based coatings with lowest wall superheat temperature. Hsu's model [24] supports in estimating these wide range of cavity sizes. The range of active nucleation cavity sizes is determined by the following equation:

$$[R_{c,max}, R_{c,min}] = \frac{\delta_t C_2}{2C_1}\left[\frac{\Delta T_{sat}}{\Delta T_{sat} + \Delta T_{sub}}\right] \times \left[1 \pm \sqrt{\frac{1 - 8C_1 \sigma T_{sat}(\Delta T_{sat} + \Delta T_{sub})}{\rho_v h_{fg} \delta_t (\Delta T_{sat})^2}}\right] \quad (3)$$

Where $C_1 = 1 + \cos\theta_r$ and $C_2 = \sin\theta_r$. $R_{c,max}$ and $R_{c,min}$ are maximum and minimum radii of the nucleation cavities, θ_r is the receding contact angle, δ_t is thermal boundary layer thickness (m), ΔT_{sat} is the wall superheat temperature ($\Delta T_{sat} = T_{surface} - T_{sat}$) (K), ΔT_{sub} is the subcooled temperature (K), σ represents the surface tension of water at saturation temperature (N/m), ρ_v is the vapor density (kg/m^3), and h_{fg} represents the latent heat of vaporization (J/kg).

Plots in **Figure** 7a and b indicate that with increase in wall superheat temperature, smaller nucleation cavities become active, providing massive enhancement in HTC due to the amplified contribution from the rapid nucleation activity. Lowest cavity diameter ranges are (**Figure** 7 a,b) observed for 3% GNP coating, which suggests the presence of more liquid supply sites than 2% and 5% GNP coating. Thus, the highest HTC is attained for 3% GNP coating. Scanning electron microscopic images in **Figure** 7**c**,**d**, and **e** indicate the different size of pores developed as a result of salt templated sintering. Wide range of pore dimensions formed on the coating assist in boiling inversion as well, leading to increment in HTC of the heater surface.

2.2.3 Dip coating technique

Dip coating, as the name suggests, is a simple deposition technique in which the heater surface is dipped in the G/GO colloidal solution for a certain duration and allowed to air dry in a controlled atmosphere after taking out from a dipping solution.

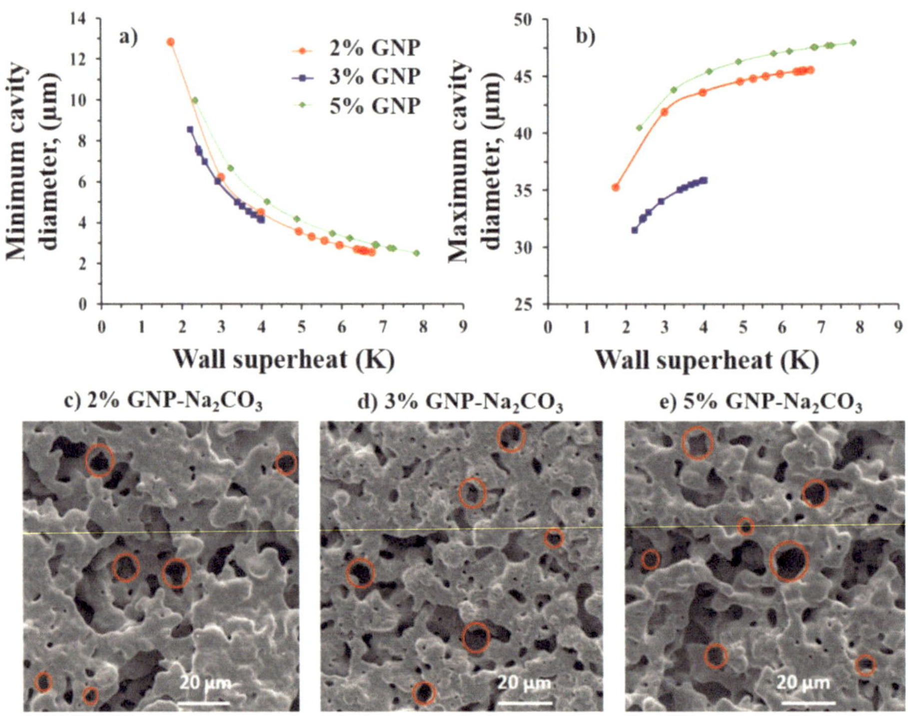

Figure 7.
Range of active nucleation cavities for 2, 3, and 5% GNP-Cu coatings showing (a) minimum cavity diameters, (b) maximum cavity diameters as a function of wall superheat temperature using Hsu's model, (c), (d), and (e) SEM images at 2kX magnification confirming the availability of wide range of porous network in the estimated range of diameters.

Different morphological features as well the thicknesses can be generated on the substrate/heater surface by varying the dipping duration.

Various techniques can be used to create a colloidal graphene solution. One of the techniques for such G/GO colloidal solution is developed using the electrodeposition technique with non-electrolyte bath. This process is developed to avoid the production of graphene using highly toxic and harmful acids, thus allowing the solution to be directly implemented for the dip coating. The electrolyte bath consists of deionized water and carbon tetrachloride 10% by volume. A potential is applied between the copper cathode and graphite anode which introduces a current density of 300 mA/cm^2 with the gap of 1 mm between the cathode and the anode. Electrodeposition technique involves cleaving of graphite electrode and reduction of the cleaved graphene oxide (GO) to form G/GO colloidal solution. The copper test surface is then dipped in the G/GO colloidal solution for a specific period of time and then is dried in a controlled atmosphere. The longer dip coating duration creates coating with less voids and fill up the copper surface with graphene, while shorter duration coating creates more ridge type structure. The microscale coating for 2 min. Dip coating attained a CHF of 1820 kW/m^2, that is $\sim$45% higher than a plain copper surface. While 10- and 20-minutes dip coating surfaces showed a slight reduction in CHF than a plain copper surface. Increased microlayer evaporation and alteration of wettability are the responsible enhancement mechanisms for improvements in pool boiling performance for 2 min. Dip coating. Longer duration coatings created less voids and thus did not assist in improving the performance [25]. This suggests that the mere presence of graphene is highly unlikely to provide any benefits in improving the pool boiling heat transfer performance and thus has very limited scope in real-world applications.

2.2.4 Concluding remarks

Some of the microscale graphene-based coating studies that have attained the highest pool boiling performance have shown that graphene does indeed play a crucial role in efficiently removing the heat from heater surfaces. It has also been shown that addition of graphene is advantageous in improving the aging and repetitive performance of the coatings.

3. Conclusions and summary

This chapter has focused on recent advances and prominent research studies that have developed graphene-based functional surfaces for enhancing the pool boiling heat transfer efficiencies of the heater surfaces. Compared to nanoscale graphene coatings, composite graphene-based coatings have shown an immense potential in increasing the overall pool boiling performance along with the longevity and sustenance of the coatings. This is because along with thermal conductivity, additional surface properties of the coatings such as wettability, wickability, and porosity are equally important and play a key role in increasing the pool boiling performance. These are essential key factors to maximize the usage of properties of the graphene in pool boiling heat transfer applications. Wide range of porous graphene-based composite wicking structures are also ideal for enhancements in heat transfer.

Amongst different composite coating techniques, both electrodeposition and sintering have shown higher cohesive and adhesive bond strengths and thus have improved longevity in maintaining the pool boiling performance. Mechanical properties of deposited graphene further assist in increasing longevity. However, additional efforts are still essential to further extend the longevity of the graphene-based composite coatings and to implement these techniques in various industrial applications. Some of the approaches include the provision of additional corrosion protection layers for composite coatings along with enhancements in bond strength and adhesion.

Acknowledgements

I would like to acknowledge my PhD advisors Dr. Satish Kandlikar and Dr. Anju Gupta for their valuable guidance.

Conflict of interest

The authors declare no conflict of interest.

Nomenclature

$c_{p,\,l}$	Specific heat of liquid, J/kg K
$c_{p,\,v}$	Specific heat of vapor, J/kg K
F	Faraday's constant
h	Heat transfer coefficient, W/m^2 K
h_{lv}	Latent heat of vaporization, J/kg
k	Thermal conductivity, W/m K
k_l	Thermal conductivity of liquid, W/m K
k_v	Thermal conductivity of vapor, W/m K
M	Atomic weight
n	Number of electrons
P_{sat}	Saturation pressure, kPa
q''	Heat flux per unit area, W/m^2
Q	Charge, C
$R_{c,min}$	Minimum radius of nucleation cavity, m
$R_{c,max}$	Maximum radius of nucleation cavity, m
T_{sat}	Saturation temperature, K
$T_{surface}$	Heater surface temperature, K

Greek letters

δ_t	Thermal boundary layer, m
$\Delta T_{sat} = T_{surface} - T_{sat}$	Wall superheat, K
ΔT_{sub}	Subcooled temperature, K
μ_v	Dynamic viscosity of vapor, Pa·s
ρ_l	Density of liquid, kg/m^3
ρ_v	Density of vapor, kg/m^3
σ	Surface tension, N/m
θ_r	Receding contact angle

Acronyms and abbreviations

APCVD	Atmospheric Pressure Chemical Vapor Deposition
CHF	Critical Heat Flux
EDS	Energy Dispersive Spectroscopy
HTC	Heat Transfer Coefficient
G	Graphene
GO	Graphene Oxide
rGO	Reduced Graphene Oxide
GNP	Graphene NanoPlatelets
Sccm	Standard cubic centimeters per minute
SEM	Scanning Electron Microscopy
rpm	Revolutions per minute

Author details

Aniket M. Rishi
LAM Research, Microsystems Engineering, Rochester Institute of Technology, Fremont, CA, USA

*Address all correspondence to: aniket.rishi@lamresearch.com; amr6756@rit.edu

References

[1] Novoselov KS, Geim AK, Morozov SV, Jiang D, Zhang Y, Dubonos SV, et al. Electric field effect in atomically thin carbon films. Science. 2004;**306**(5696):666-669

[2] Jang M. The potential power of graphene. International Journal of Students' Research. 2016;**1454**(3): 1-12

[3] Jorio A, Dresselhaus MS, Saito R, Dresselhaus G. The Sp2 Nanocarbons: Prototypes for nanoscience and nanotechnology. In: Raman Spectroscopy in Graphene Related Systems. Weinheim, Germany: Wiley-VCH Verlag GmbH & Co. KGaA; 2011. pp. 1-15

[4] Chen Y-M, He S-M, Huang C-H, Huang C-C, Shih W-P, Chu C-L, et al. Ultra-large suspended graphene as a highly elastic membrane for capacitive pressure sensors. Nanoscale. 2016;**8**(6): 3555-3564

[5] Gupta A, Jaikumar A, Kandlikar SG, Rishi A, Layman A. A multiscale morphological insight into graphene based coatings for Pool boiling applications. Heat Transfer Engineering. 2018;**39**(15):1331-1343

[6] Mudhafar MAH, Jhang B-W, Sin Y-Y. Experimental comparison of Pool boiling performance between Cu-plain surface and functionalized graphene layers. International Journal of Refrigeration. 2022;**139**:1-12

[7] Choi G, Yun M, Hsu W-T, Shim DI, Lee D, Kim BS, et al. Enhanced boiling heat transfer by nucleation patterning with self-assembly of reduced graphene oxide coating. International Journal of Heat and Mass Transfer. 2022; **197**(123329):1-10

[8] Choi G, Shim DI, Lee D, Kim BS, Cho HH. Enhanced nucleate boiling using a reduced graphene oxide-coated micropillar. International Communications in Heat and Mass Transfer. 2019;**109**(104331):1-8

[9] Mukalil SK, Gajghate SS, Majumder A, Saha BB, Bhaumik S. Numerical and experimental analysis of Pool boiling heat transfer using graphene Nanofluid. Proceedings of International Exchange and Innovation Conference on Engineering & Sciences (IEICES), Kyushu University, Japan. 2020;**6**: 264-270

[10] Ahn HS, Kim JM, Kaviany M, Kim MH. Pool boiling experiments in reduced graphene oxide colloids part II – Behavior after the CHF, and boiling hysteresis. International Journal of Heat and Mass Transfer. 2014;**78**:224-231

[11] Vasudevan D, Senthilkumar D, Surendhiran S. Performance and characterization studies of reduced graphene oxides aqua Nanofluids for a Pool boiling surface. International Journal of Thermophysics. 2020;**41**(6): 41-74

[12] Kandlikar SG. Handbook of Phase Change: Boiling and Condensation. 1st ed. New York, USA: Routledge; 1999. 784 pages

[13] Pioro IL, Duffey RB. Heat Transfer & Hydraulic Resistance at Supercritical Pressures in Power Engineering Applications. New York, NY, USA: ASME Press; 2007. 334 pages

[14] Kim JH, Simon TW, Viskanta R. Journal of heat transfer policy on reporting uncertainties in experimental measurements and results. Journal of Heat Transfer. 1993;**115**(1):5-6

[15] Ganesan UK, Soni K, Suresh S, Ghosh K, Thansekhar MR, Dinesh Babu P. Modified surfaces using seamless graphene/carbon nanotubes based nanostructures for enhancing Pool boiling heat transfer. Experimental Thermal and Fluid Science. 2018;**96**: 493-506

[16] Hu Y, Li H, He Y, Wang L. Role of nanoparticles on boiling heat transfer performance of ethylene glycol aqueous solution based graphene Nanosheets Nanofluid. International Journal of Heat and Mass Transfer. 2016;**96**:565-572

[17] Barker D, Walsh FC. Applications of Faraday's Laws of electrolysis in metal finishing. Transactions of the Institute of the Metal Finishing. 1991;**69**(4):158-162

[18] Jaikumar A, Rishi A, Gupta A, Kandlikar SG. Microscale morphology effects of copper–graphene oxide coatings on Pool boiling characteristics. Journal of Heat Transfer. 2017;**139**(11): 111509-111520

[19] Rishi AM, Kandlikar SG, Gupta A. Improved wettability of graphene Nanoplatelets (GNP)/copper porous coatings for dramatic improvements in Pool boiling heat transfer. International Journal of Heat and Mass Transfer. 2019; **132**:462-472

[20] Ahmadi VE, Khaksaran MH, Apak AM, Apak A, Parlak M, Tastan U, et al. Graphene-coated sintered porous copper surfaces for boiling heat transfer enhancement. Carbon Trends. 2022; **8**(100171):1-10

[21] Rishi AM, Kandlikar SG, Rozati SA, Gupta A. Effect of ball milled and sintered graphene Nanoplatelets–copper composite coatings on bubble dynamics and Pool boiling heat transfer. Advanced Engineering Materials. 2020;**22**(7):1-11, 1901562

[22] Rishi AM, Kandlikar SG, Gupta A. Salt Templated and graphene Nanoplatelets draped copper (GNP-draped-Cu) composites for dramatic improvements in Pool boiling heat transfer. Scientific Reports. 2020;**10**(1): 1-14. 11941

[23] Majumder B, Pingale AD, Katarkar AS, Belgamwar SU, Bhaumik S. Enhancement of Pool boiling heat transfer performance of R-134a on microporous Al@GNPs composite coatings. International Journal of Thermophysics. 2022;**43**(4):1-19. 49

[24] Hsu YY. On the size range of active nucleation cavities on a heating surface. Journal of Heat Transfer. 1962;**84**(3): 207-213

[25] Jaikumar A, Gupta A, Kandlikar SG, Yang C-Y, Su C-Y. Scale effects of graphene and graphene oxide coatings on Pool boiling enhancement mechanisms. International Journal of Heat and Mass Transfer. 2017;**109**: 357-366

Chapter 4

Boiling and Condensation in Two-Phase System Transients with Water Hammer

Sanja Milivojevic, Vladimir Stevanovic, Milan M. Petrovic and Milica Ilic

Abstract

Water hammer in two-phase systems, induced by direct steam condensation on subcooled water or by separation of subcooled water column, results in the most intensive pipeline pressure surges. Amplitudes of pressure spikes along the course of these dangerous transients strongly depend on the condensation and evaporation rates. The present paper provides a literature overview of thermal-hydraulic models for the prediction of water hammer phenomenon in two-phase systems, together with an original mechanistic approach for the prediction of phase transition rates, based on the shape and size of vapor-liquid interfacial area and the phase transition potential expressed through vapor and liquid phase temperature difference. Available water hammer experimental conditions were numerically simulated with the new modeling approach. Driving parameters of boiling and condensation rates at the steam-water interfaces are evaluated, and a good agreement is shown between numerical results and experimental data of bulk two-phase flow parameters during water hammer transients.

Keywords: water hammer, two-phase flow, steam, condensation, modeling, numerical simulation

1. Introduction

The water hammer is recognized as a very dangerous phenomenon, and therefore, the prediction of its occurrence is necessary in order to prevent accidents. Depending on the mechanisms of their origin, we can divide water hammers into two types [1]. One type of water hammer, in single-phase flow, is caused by a quick valve opening or closure or when a pump suddenly stops. During the course of a transient with rapid flow rate changes, a column separation might occur, which is characterized with the rapid gaseous and vaporous cavitation and the formation of a two-phase system [2]. The other type of water hammer is caused by rapid condensation of steam in direct contact with subcooled liquid in a pipe or vessel under pressure. This second type is called condensation induced water hammer (CIWH). The CIWH and water hammer with column separation take place in two-phase system with rapid boiling and condensation. An overview of investigation of these two mechanisms of water hammers in two-phase systems follows.

1.1 Water hammer with column separation

The safety of various hydraulic systems depends on the accuracy of the prediction of water hammer with gaseous and vaporous cavitation. The first pressure surge in the single-phase system, caused by a sudden valve closure, causes a maximum peak pressure, close to the pressure rise in liquid obtained by the Joukowsky relation,$\Delta p = -\rho c \Delta u$, where ρ is the fluid density, c is the sonic velocity in the fluid-pipe system, and Δu is the fluid velocity change. The intensities of the subsequent pressure peaks during the water hammer transient are greatly affected by the presence of dissolved air in the liquid and vapor generation due to the evaporation of liquid if pressure is lower or equal to the saturation pressure. In all systems where water comes into contact with air, the air dissolves in the water so that water contains entrained air microbubbles. The air content in untreated tap water is 1.13×10^9 microbubbles per m^3. The most probable air microbubble diameter is 6×10^{-6} m. In degassed water, the air content is reduced to 0.911×10^9 microbubbles per m^3 with the same most probable microbubble diameter [3]. A detailed review of solubility of air and solubility of other gases in water can be found in [4]. Before the occurrence of the water hammer, entrained air microbubbles have the same velocity as a water volume and practically no influence on the thermophysical parameters of air-water mixture. If during the pressure transient pressure drops and water is degassed, a significant influence on the hydrodynamics of the mixture of water and air is observed. In cases with pressure drops below the saturation pressure along the course of the transient, the liquid becomes superheated, and adiabatic evaporation occurs, i.e., bubbles are generated although no heat is added to the liquid. The main difference between the two-phase mixture patterns with adiabatic evaporation and diabatic wall boiling is in the locations of bubbles nucleation and rise and corresponding void distribution. In case of adiabatic evaporation, the nucleation of bubbles occurs both on the heated microscopically rough wall surface and on the impurities within the bulk of liquid phase, i.e., a rapid bubbling or so-called flashing occurs within the whole liquid volume, while in case of wall boiling, the liquid is superheated within the thin liquid layer on the heated wall, and the bubbles are mainly generated on the wall surface and hydrodynamically transferred to the bulk of liquid volume [5]. Therefore, in cases of wall boiling, the void fraction is mainly higher in the vicinity of wall surface, which especially holds in cases of subcooled boiling with bubbles condensation within the bulk of subcooled liquid.

Bergant and Simpson [6] compared results of several numerical models with the data measured within an experimental test of the water hammer caused by the rapid valve closure. The system is described with a set of one-dimensional equations. This set includes water hammer equations for single liquid flow, two-phase flow equations for a distributed vaporous cavitation region, shock equations for condensation of liquid-vapor mixture back to the liquid and equations for a discrete vapor cavity separating a liquid and a vaporous cavitation region. The occurrence of cavitation during the transient was simulated using three models: the discrete vapor cavitation model (DVCM), the discrete gas cavitation model (DGCM) and the generalized interface vapor cavitation model (GIVCM). In the DVCM, it is assumed that the location of the gas phase is at the grid nodes, the liquid phase fills the space between nodes, the speed of wave propagation between adjacent nodes is equal to the sonic velocity in the liquid phase, the minimum pressure during the transient is determined by the pressure at which the first vapor bubble forms (liquid/vapor saturation pressure at the liquid temperature), and there is no wave propagation in the two-phase flow [7]. The model is applied to discrete cavities in which vapor bubbles as well as vapor cavities are placed. The DVCM is the most widely used for modeling vapor cavitation during transient in hydraulic systems, but its accuracy depends on the ratio of the cavitation volume and the volume between the numerical nodes, which is its main drawback [8]. The DGCM model is simple and gives good results for a wide range of input parameters if the gas fraction in the working fluid is small [9]. It is fully specified with its characteristic equations, the continuity equation for the ith numerical node and the ideal gas law. DGCM has been successfully used to model both vapor and gas cavitation. The GIVCM explicitly describes discrete cavities and vapor cavitation regions. As a basis for the development of the cavitation vapor interface, the algorithm of the DVCM was used, which allows the cavities to be formed into a branched network, which is calculated using the method of characteristics (MOC). The weakness of a model, in comparison with DVCM, is a long computational time. The main difference between the three models used is in the physical interpretation and description of the formation of the vapor cavitation region that spreads from the middle cavity towards the closed valve.

The limitations of DVCM, DGCM and GIVCM cavitation models, which are a consequence of adopted assumptions, are overcome in the presented mechanistic modeling approach, based on non-equilibrium gas release and absorption as well as on vaporous cavitation. Previous research has shown the lack of data on the air content in the liquid during the pressure transient and that it is necessary to explain the gas release in more detail. In this book chapter, the homogeneous gas-liquid two-phase flow model is applied to the simulation of water hammer with gaseous and vaporous cavitation, which includes non-equilibrium gas-liquid and vapor-liquid mass transfer at the interface.

1.2 Condensation induced water hammer

Direct contact of steam and subcooled water leads to CIWH. Since the specific volume of steam is significantly greater than the specific volume of liquid, there is a pressure drop in a part of the pipe occupied by the steam. The pressure difference in the parts of the pipe occupied by the liquid phase and the parts occupied by the vapor causes the liquid column movement and acceleration towards the area occupied by the vapor. Consequently, the vapor condensation continues towards the propagating liquid and vapor interface. A liquid column accelerates and eventually hits the obstacle, such as a valve, the closed end of a pipe or another liquid column, and reflects from the obstacle. A sudden pressure pulse, generated at the moment of impact of a liquid column onto an obstacle and the consequent propagation of pressure waves, can cause severe mechanical damage to the equipment, such as damage to pipe walls, fittings and hangers or pressure vessels, and might endanger the safety and cause serious injury to operating personnel. The dangerous effects of CIWH were shown by Milivojevic et al. [10] with simulations of destructive pressure peaks greater than 10 MPa in systems that were initially at a low pressure close to the atmospheric.

With a more precise insight into the mechanism of the occurrence of water hammer, it is possible to improve the protection systems, implement necessary safety measures and thus prevent its consequences. Some of the facilities where this undesirable thermal–hydraulic phenomenon can occur are steam power plant units [11], nuclear power plants (NPPs) [12–14], district heating systems [15] and the ammonia refrigeration system [16]. In the previous research, whether experimental or numerical, the goal was mostly to record the highest peak pressure value due to safety reasons. An experimental and analytical investigation was performed to estimate the impulse generated during the large steam bubble collapse in a vertical pipe between the lower stagnant hot water column and the upper downward accelerating column of cold water [17]. The liquid column was observed as a rigid body, and a simple mechanical model was derived for the prediction of the water column velocity and impulse at the moment of impact and the resulting pressure peak. The CIWH in a vertical pipe filled with steam and closed at the top was reported in [18]. In this study, the pipe filled with steam is immersed in the reservoir of subcooled water. The direct contact of steam and subcooled water is caused by fast opening valve at the bottom of the pipe. The experiments showed cases with great pressure pulses, from initial several bars to approximately 16 MPa. These experimental conditions were simulated with the thermal-hydraulic nuclear reactor safety code TUF [12]. The CIWH simulations required the TUF code upgrading with the model for the steam-water interfacial area concentration at the water column head. It was assumed that the water column and steam interface consist of certain bubbles and droplets, which are formed during the interface movement. This interfacial area was kept constant during the water column movement.

The safety of NPP steam generators and feedwater systems is of great importance. Consequently, CIWH caused by countercurrent flow of steam and subcooled water in the horizontal pipe, or in pipes with small horizontal inclination, is most common, and also the most experimentally investigated in the latest decade [1, 2, 19–22], as well as earlier [23–25]. In such systems, stratified flow occurs, followed by steam bubbles formation and the rapid condensation of steam bubble entrapped by the subcooled water. Some guidelines are given for the prevention of CIWH in a two-phase flow. Barna and Ezsol [25] performed the water hammer experiment in the countercurrent flow of steam and water in the horizontal pipe in the Hungarian PMK-2 facility and also in the Japanese Rig-of-safety Assessment (ROSA) facility. In addition, they performed numerical simulation by WAHA3 code for transients in single- and two-phase flow. Large system codes like RELAP5, Trac and CATHARE are used for the safety analysis of transients in NPP, and they model two-phase flow transients. Numerical simulations of CIWH are performed by using large system codes as ATHLET, WAHA3 [25], RELAP5 [20], Trac, CATHARE, in-house codes [26] and OpenFOAM, open source CFD software [21]. A type of the CIWH known as the water cannon phenomenon can occur during the discharge of steam into a large volume of cold water. In case when the exhaust valve is closed, steam can be found trapped in a pipeline. The rapid condensation of steam on cold water is followed by subcooled water suction into the pipe. The formed water slug eventually hits the valve and causes a large pressure pulse. Yeung et al. [27] performed a simulation of the water cannon phenomenon by nuclear reactor safety code RELAP5/MOD3. Dynamics of CIWH depends on complex thermal and hydraulic effects, such as the steam condensation on subcooled water, the disintegration of the water column head and the droplets entrainment from the water column head to the steam space and the transient friction of the water column. Experimental investigations have shown that practically the same experimental conditions can result in significant scattering of the experimental test data, as it was reported in [12, 22, 28]. This scattering of the results is caused by the stochastic nature of the water column head disintegration. The steam–cold water contact area affects the way the condensation at the water column head develops which further depend on the acceleration of the water column and the peak pressure at the impact during CIWH. The entrainment of water droplets from the accelerated water column head and the condensation of steam onto the droplets and the disturbed water column head were investigated experimentally in [29]. The Taylor instability of the accelerated water column head was observed. The relative velocity of the entrained droplets and the water column head was correlated with the column head acceleration.

Hou et al. [26] experimentally and numerically investigated flashing instability induced water hammer (FIIWH), which belongs to the class of CIWH, and occurs in open natural circulation systems (NCS). The occurrence of FIIWH is more likely in NCS with long horizontal pipes. NCS are widely used in NPPs as the components of the passive safety systems. Sun et al. [19] experimentally investigated CIWH in NCS, regarding flow parameters and fluid temperature. Since they used tap water as a medium, in which there are dissolved non-condensable gases, when a CIWH occurred, these gases were released. The volume of non-condensable gases did not change during the experiment, which shows that all the vapor in the bubbles, which occurred after CIWH, completely evaporated. The presence of non-condensable gases in the entrapped bubble reduces the heat transfer coefficient and lowers condensation rate, so that the speed of the water columns is lower, which results in smaller pressure peaks when they collide. A prerequisite for the formation of CIWH is the existence of an entrapped bubble, and the degree of subcooling of the water determines whether or not a CIWH will occur. The steam in the entrapped bubble will condense in contact with surrounding subcooled water, the pressure in the bubble will suddenly drop, then the water at the bottom of the bubble will instantly evaporate, and steam condensation induced flashing (SCIF) will occur. The SCIF occurs when water on one side of the entrapped bubble is saturated and on the other is subcooled. The SCIF diminishes the pressure drop in the bubble due to condensation and slows down the occurrence of CIWH. After CIWH in NCS water column moves in the opposite direction of the initial flow. Sun et al. [19] conducted 67 CIWH events, and they identified three types of CIWH in NCS. Type I CIWH occurs when there is a large difference in velocity of water and steam, slip ratio. The velocity change results in pressure change which increases the disturbances at the interface. When pressure change prevails over gravity and surface tension, Kelvin-Helmholz instability occurs. Type II CIWH is caused by the solitary wave formation and reverse flow of subcooled water into the pipe. Interaction of solitary and interface wave is also important. Type III CIWH is induced by the pressure wave generated after the occurrence of CIWH. The most probable type is Type II CIWH, and the least probable is Type III.

The transient friction strongly affects water column acceleration during CIWH, especially in the vicinity of the pressure wave front, because it affects the propagation speed of the wave front and the evolution of its shape. By modeling vapor cavitation in transient fluid flow, Shu [30] concluded that by applying the unsteady friction model, better damping is achieved, in cases where the cavitation is not too strong and when the assumption of laminar single-phase flow with the appropriate weighting function is applicable. An unsteady turbulent skin friction model for one-dimensional smooth pipes, where the wall shear stress is proportional to the fluid instantaneous acceleration, is developed by Vardy and Brown [31]. Errors resulting from the numerical simulation of friction in unsteady flows in small pipe networks were evaluated in [32] in the case of applying the MOC with a fixed node distance in the numerical grid and without interpolation. The application of the friction relaxation model to the prediction of the water hammer in two-phase flow is presented in [33].

The dynamics of deaeration and the fluid structure interaction (FSI) after the occurrence of the first pressure pulse are also effects that are distinguished as very influential on CIWH. Neuhaus and Dudlik in [34] experimentally investigated the air release process. The FSI model was implemented into fluid dynamic equations. Taking into account the air release as well as the FSI model predictions are improved.

In order to improve the process safety and to prevent equipment damage and accidents, it is very important to single out the thermal–hydraulic conditions that can lead to CIWH, as well as the maximum pressure peaks and periods of generated pressure pulses during CIWH. The dynamics of generated pressure pulses during CIWH event is predominantly governed by the intensity of the condensation rate. Direct contact of subcooled liquid and steam leads to intensive condensation. The condensation rate is highly influenced by the interfacial area concentration and condensation heat transfer coefficient. The interfacial area has a very irregular shape. During CIWH transient liquid jets are formed, and liquid column is disintegrated so that droplets from the head are entrained in steam volume. This droplet entrainment increases interfacial area and condensation rate. In the available literature, there is a lack of information on the determination of the condensation rate. Liu presented in [12] model of condensation during CIWH event, but without using values for interfacial area concentrations of steam and liquid droplets and also of liquid and steam bubbles in close to the water column head. In studies by [27, 28, 33] the rate of condensation was only mentioned without providing further details. Yeung et al. [27] referred that a combination of mechanistic models and experimental correlations was used in the calculation using the RELAP code. Barna et al. [28] mentioned the quick condensation model in WAHA code, but no detailed information was provided. Kucienska et al. [33] outlined the use of the heat and mass transfer model for dispersed flow. The model is derived from the homogeneous relaxation approach, assuming a large heat transfer coefficient at the vapor side of phase interface. No further details about the model application are given.

The next chapter presents our own numerical model for the prediction of CIWH and for the prediction of water hammer with gaseous and vaporous cavitation.

2. Modeling approach

2.1 Governing equations

Single-phase vapor or liquid and two-phase vapor-liquid flow in a pipe is considered as the one-dimensional, transient and compressible fluid flow of homogeneous fluid. The velocity and thermal equilibrium are assumed between the vapor and liquid phase in the two-phase flow. The following mass, momentum and energy conservation equations are applied:

- Mass conservation

$$\frac{D\rho}{Dt} + \rho \frac{\partial u}{\partial x} = 0 \tag{1}$$

- Momentum conservation

$$\frac{Du}{Dt} + \frac{1}{\rho}\frac{\partial p}{\partial x} + \frac{fu|u|}{2d_{\mathrm{H}}} + f_{\mathrm{u}}\mathrm{sign}(u)\left|\frac{\partial u}{\partial t}\right| + g\sin\theta = 0 \tag{2}$$

- Energy conservation

$$\frac{Dh}{Dt}-\frac{1}{\rho}\frac{\mathrm{D}p}{\mathrm{D}t}-\frac{fu^2|u|}{2d_\mathrm{H}}-f_\mathrm{u}u\mathrm{sign}(u)\left|\frac{\partial u}{\partial t}\right|+\int_{x_\mathrm{i}-\varepsilon}^{x_\mathrm{i}+\varepsilon} sign\left(h-h'\right)\frac{\Gamma_c h''}{\rho}dx=0 \tag{3}$$

where the dependent variables are velocity u, pressure p and specific enthalpy h, and independent variables are time t and spatial coordinate x. The hydraulic diameter is denoted with d_H, the pipe inclination angle from the horizontal axis with θ, gravity acceleration with g, the Darcy friction coefficient with f and the unsteady friction coefficient with f_u. The third and fourth terms on the left-hand side of momentum and energy conservation equations, Eqs. (2) and (3), are related to pressure drop due to steady-state and transient friction, respectively. The last term on the left-hand side of the energy Eq. (3) determines interfacial heat transfer per unit mass of fluid due to condensation.

2.2 Closure laws for condensation induced water hammer

The pressure drop due to friction is determined using Darcy friction coefficient f. The unsteady friction coefficient is determined according to [31], as

$$f_\mathrm{u}=2\sqrt{\frac{12.86}{\mathbf{Re}^{\left(1.1844-0.0567\log_{10}\mathbf{Re}\right)}}} \tag{4}$$

where **Re** is the Reynolds number.

The condensation rate in Eq. (3) is determined as

$$\Gamma_c=\frac{q_c a_i}{r} \tag{5}$$

where r is the latent heat of condensation, and the condensation heat flux q_c is determined as the product of the condensation heat transfer coefficient h_c and the difference between the saturation temperature T_{sat} and the subcooled liquid temperature T_1

$$q_c=h_c(T_{sat}-T_1). \tag{6}$$

The condensation takes place from vapor or vapor-liquid two-phase mixture to the liquid column at the interface, whose position is denoted with x_i (**Figure 1**). Parameter ε represents infinitesimal distance from the interface to the subcooled liquid and from the interface to the two-phase mixture or vapor. The transfer of thermal energy of condensation through the interface from the two-phase mixture or vapor to the liquid column is determined by the function $sign\left(h-h'\right)$.

The prediction of interfacial area concentration a_i is the main task when determining the condensation rate, besides determining the condensation heat transfer coefficient. Throughout the CIWH event transient liquid droplets or jets separate from the liquid column head and entrain in the vapor volume, so that the interfacial area increases, as well as the condensation rate. The developed model assumes that disintegration process of liquid column head, the entrainment of droplets into vapor

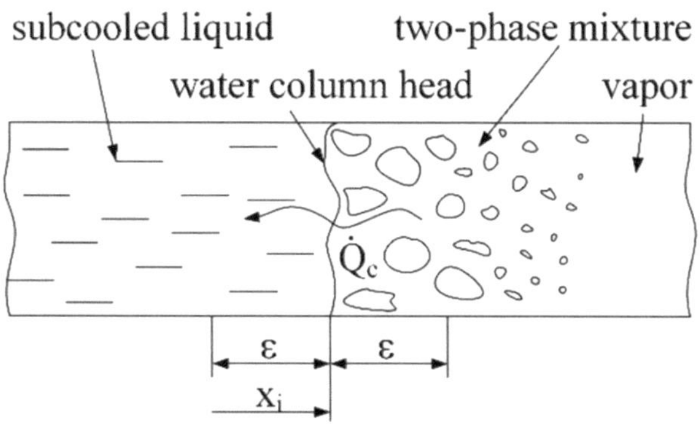

Figure 1.
Vapor-liquid interface position and the direction of the condensation process.

and the intensity of the heat transfer at the interface depend on the liquid column head acceleration. The product of the condensation heat transfer coefficient and the interfacial area concentration is divided to two additions, first related to liquid column head (LCH) and the other to entrained droplets (ED)

$$h_c a_i = (h_c a_i)_{LCH} + (h_c a_i)_{ED}. \tag{7}$$

Bloemeling [35] used correlations based on surface renewal theory to predict the condensation of steam at the turbulent liquid column head, but greater uncertainty occurs in predicting turbulent characteristic length and velocity. These turbulent scales are correlated with turbulent kinetic energy and dissipation rate, and accuracy of their calculations requires application of 3D models of steam–water interface behavior in transient conditions. In this chapter, the condensation heat transfer coefficient at the LCH is calculated using Dittus and Boelter (1930) correlation, i.e., turbulent heat transfer at the LCH is considered as convective heat transfer at the pipe wall

$$(h_c)_{LCH} = (\lambda_1/d_H)\, 0.023\, \mathrm{Re}_1^{0.8}\, \mathrm{Pr}_1^{0.4} \tag{8}$$

where index 1 denotes liquid, λ is the thermal conductivity, and **Re** and **Pr** are Reynolds and Prandtl number, respectively. The interfacial area concentration at the LCH is calculated with the assumption that it is equal to the cross section of the pipe

$$(a_i)_{LCH} = \frac{A_i}{A_i \Delta x} = \frac{1}{\Delta x}. \tag{9}$$

The condensation heat transfer coefficient on the ED is determined from the Nusselt number, whose constant value is theoretically predicted and experimentally confirmed [36–38] as

$$\mathrm{Nu} = \frac{(h_c)_{ED} d_D}{\lambda_1} = C. \tag{10}$$

The adopted constant value for C is 17.9 according to Kronig and Brink [36]. The entrained droplet diameter depends on parameter Y

$$Y = \frac{\sigma \mathrm{We}_{cr}}{\rho_2 (u_2 - u_1)^2} \tag{11}$$

where σ is the surface tension, index 2 denotes superheated vapor, and the value of the critical Weber number $\mathbf{We_{cr}}$ = 0.799 is taken from [39]. The entrained droplet diameter is calculated as

$$d_D = \begin{cases} 10^{-4}\mathrm{m}, & Y \leq 10^{-4}\mathrm{m} \\ Y, & 10^{-4}\mathrm{m} < Y < 5 \cdot 10^{-4}\,\mathrm{m}. \\ 5 \cdot 10^{-4}\mathrm{m}, & Y \geq 5 \cdot 10^{-4}\mathrm{m} \end{cases} \tag{12}$$

The major challenge here exists in determination of the interfacial area concentration between entrained droplets from the liquid column head and the vapor. It is assumed that the interfacial area concentration depends on the acceleration of the liquid column head as follows

$$(a_i)_{ED} = a_{D,0} + 4 \cdot 10^{-3} (Du/Dt)^3 \tag{13}$$

Where $a_{D,0}$ takes the values between 0 and 40 m^2/m^3. The determination of the parameters on the right-hand side of Eq. (13) is performed by comparing the results of numerical simulations of CIWH with experimental data available in [12, 18] and data in [27]. The term $a_{D,0}$ is determined by dynamics of propagation of the liquid column towards the vapor volume, and it depends on the initial conditions under which CIWH occurs, e.g., the fast opening of the valve or removing the obstacle between superheated liquid and vapor. The formation of the interfacial area between liquid and steam during the opening of the valve is stochastic in nature. Interfacial area is very irregular with entrained droplets in vapor and formation of liquid jets, and therefore, $a_{D,0}$ is in the range between 0 and 40 m^2/m^3. If its value is 0, that means that there are no entrained droplets which significantly increase the interfacial area. The second term in Eq. (13) takes into account the influence of the acceleration of the liquid column head on the disintegration of the column head and the droplet entrainment in the vapor volume.

The thermodynamic quality is used to determine the phase state of the fluid

$$x_t = \frac{h - h'}{r} \tag{14}$$

where for $x_t \leq 0$ liquid phase takes place, for $x_t \geq 0$ vapor phase, and for $0 < x_t < 1$ a two-phase mixture. The fluid density is calculated as a reciprocal value of the specific volume determined by applying the equations of state for subcooled liquid $v_1(p,h)$, superheated vapor $v_2(p,h)$, saturated liquid $v'(p)$ and saturated vapor $v''(p)$ (data from the steam tables [40] are used)

$$\rho = \begin{cases} 1/v_1(p,h), & \text{if } x_t \leq 0 \\ 1/(v'(p) + x_t(v''(p) - v'(p))), & \text{if } 0 < x_t < 1. \\ 1/v_2(p,h), & \text{if } x_t \geq 1 \end{cases} \tag{15}$$

2.3 Closure laws for water hammer with gaseous and vaporous cavitation

Liquid single-phase and gas/vapor-liquid two-phase flow in a pipe is observed as the one-dimensional, transient and compressible fluid flow of homogeneous fluid. The velocity equilibrium is assumed between the gas/vapor and liquid phase in the two-phase flow. The flow is isenthalpic. Mass and momentum conservation equations Eqs. (1) and (2) are applied, where the unsteady friction coefficient is calculated according to Eq. (4) and the fluid density is evaluated by Eq. (15).

The reasonable assumption is that the liquid density is constant and determined by the initial liquid temperature and the initial average pressure in the flow channel. The gas/vapor density is calculated with the ideal gas law

$$\rho_2 = \frac{p_2}{R_g T_2} \tag{16}$$

where T_2 is the gas/vapor temperature approximated with the initial liquid temperature $T_{1,0}$, while R_g is the gas constant. In the case of bubbly flow, the gas/vapor pressure is calculated by taking into account the surface tension σ, according to the Laplace eq. [38]

$$p_2 = p_1 + \frac{2\sigma}{r_b}. \tag{17}$$

Assuming that there is no heat exchange between the flow channel and the surroundings and that the single-phase flow and homogeneous two-phase flow are isenthalpic, gas/vapor density only depends on pressure

$$\rho_2 = \rho_2(p_2). \tag{18}$$

The infinitesimal change in gas/vapor quality in the case of water hammer with gaseous or vaporous cavitation is determined as [41–43].

$$\frac{dx}{dt} = \frac{\Gamma}{\rho} \tag{19}$$

where Γ represents the rate of interfacial mass transfer of gas/vapor (due to evaporation and condensation in case of vaporous cavitation and due to absorption and desorption in case of non-condensable gaseous cavitation) per unit volume and per unit of time, and ρ is the gas/vapor-liquid two-phase mixture density. The rate of interfacial mass transfer of gas/vapor is determined as

$$\Gamma = j_i a_i \tag{20}$$

where a_i is the gas/vapor-liquid interfacial area concentration, and j_i is the interfacial mass transfer flux of vapor or non-condensable gas.

The relation between the gas/vapor void fraction and quality is [38].

$$\alpha_2 = \frac{1}{1 + \frac{1-x}{x}\frac{\rho_2}{\rho_1}}. \tag{21}$$

The calculation of gas/vapor-liquid interfacial area concentration depends on gas/vapor quality and the two-phase flow pattern. It is assumed that for voids lower than or equal to 0.74 the flow pattern is bubbly [44], while for voids greater than 0.74 the annular pattern is assumed. The interfacial area concentration in the bubbly flow is calculated as [37].

$$a_i = \frac{6\alpha_2}{d_b} \tag{22}$$

with the assumption that the bubbles are spheres with uniform diameter d_b. The bubble diameter is calculated from the relation

$$\alpha_2 = \frac{n_b \pi d_b^3}{6} \tag{23}$$

where the gas/vapor void fraction is related to the bubble concentration n_b. The interfacial area concentration for the annular flow pattern is calculated from the following expression by assuming that the liquid phase wets the tube wall as liquid film [37].

$$a_i = \frac{4}{d}\sqrt{\alpha_2} \tag{24}$$

where d is the tube inner diameter.

The interfacial mass transfer flux of vapor on the surface of bubble, in case of vaporous cavitation, is

$$\dot{j}_i = \rho_2 \frac{dr_b}{dt}. \tag{25}$$

where ρ_2 denotes vapor density, and r_b is the bubble radius. The time change of the bubble radius is described by the correlation [45].

$$\frac{dr_b}{dt} = \frac{\lambda_1}{\rho_2 r}\left(\frac{1}{(a_1 t)^{1/2}} + \frac{1}{r_b}\right)(T_1 - T_{sat}). \tag{26}$$

where λ_1 denotes thermal conductivity of liquid, r is latent heat, a_1 is the thermal diffusivity of liquid, T_1 and T_{sat} are, respectively, liquid temperature and saturation temperature. During the bubble growth, the bubble radius r_b is much greater than value of $(a_1 t)^{1/2}$so that the equation for time change of the bubble radius can be reduced to

$$\frac{dr_b}{dt} = \frac{\lambda_1}{\rho_2 r (a_1 t)^{1/2}}(T_1 - T_{sat}). \tag{27}$$

The interfacial mass transfer rate of vapor in bubbly flow is determined by introducing Eq. (25) and Eq. (22) into Eq. (20) and then introducing the bubble diameter from Eq. (23) and the time change of the bubble radius from Eq. (27) into obtained relation. The following expression is derived

$$\Gamma = 4.835\frac{\alpha_2^{2/3} n_b^{1/3} \lambda_1}{r(a_1 t)^{1/2}}(T_1 - T_{sat}). \tag{28}$$

In the case of annular flow, the evaporation and condensation interfacial mass flux at the liquid film surface is determined from the surface heat flux equation for heat conduction in semi-infinite solid with the initial temperature T_1, that is suddenly lowered and maintained at a temperature T_{sat} as

$$q = \frac{\lambda_1}{(\pi a_1 t)^{1/2}}(T_1 - T_{sat}). \tag{29}$$

By dividing Eq. (29) with latent heat, the evaporation and condensation interfacial mass flux is obtained

$$j_i = \frac{\lambda_1}{r(\pi a_1 t)^{1/2}}(T_1 - T_{sat}). \tag{30}$$

The interfacial mass transfer rate of vapor in annular flow is determined by introducing Eq. (24) and Eq. (30) into Eq. (20) as

$$\Gamma = \frac{4\lambda_1 \alpha_2^{1/2}}{rd(\pi a_1 t)^{1/2}}(T_1 - T_{sat}). \tag{31}$$

The absorption and desorption rate of non-condensable gas in liquid, in case of gaseous cavitation, is determined in the following manner. If non-condensable gas is in contact with liquid for a long time period, the liquid becomes saturated with gas, and the equilibrium condition is reached. The molar fraction of absorbed non-condensable gas in liquid $\tilde{x}_{eq}$ is then determined by Henry's law [4].

$$\tilde{x}_{eq} = \frac{p_2}{H_c} \tag{32}$$

where p_2 is non-condensable gas pressure, and H_c is Henry's constant, which depends on the type of gas, as well as the pressure and temperature of the liquid.

Under non-equilibrium conditions, when the mass fraction of the dissolved gas in the liquid is different from the equilibrium gas saturation in the liquid, the gas transfer occurs at the interface between the liquid and gas phase. Assuming that the relative velocity between the gas and liquid phase is negligible in the case of water hammer, the convective mechanisms of gas transfer are neglected, and interface gas transfer is determined by the gas diffusion on the liquid side of the interface. In bubbly flow, which occurs during water hammer with gas cavitation, small bubbles are dispersed in the liquid, gas–liquid relative velocity can be neglected. The mass balance equation of the non-condensable gas in liquid in the coordinate system connected with the moving boundary, for isothermal absorption or desorption [46], is

$$\frac{\partial C}{\partial t} + u_i \frac{\partial C}{\partial x} = D_1 \frac{\partial^2 C}{\partial x^2} \tag{33}$$

where C is the mass fraction of dissolved non-condensable gas in liquid, D_1 is the liquid diffusivity, x is the coordinate, and u_i is the boundary displacement velocity defined by the condition of interface impenetrability for the liquid phase

$$j_{i,1} = (1 - C_i)\rho_1 u_i - \rho D_1 \frac{\partial(1 - C)}{\partial x} = 0 \tag{34}$$

as

$$u_i = -\frac{D_1}{1 - C_i}\frac{\partial C}{\partial x} \text{ (at x = 0)} \tag{35}$$

and $j_{i,1}$ in Eq. (34) is the mass flux of the liquid phase which contains convection and diffusion components. The solution to Eq. (33), for boundary displacement velocity defined by Eq. (35), with following boundary conditions: if $x = 0, C = C_i$ and when $x \to \infty$, $C = C_1$ is

$$\left.\frac{\partial C}{\partial x}\right|_{x=0} = -\frac{(C_i - C_1)}{\sqrt{\pi D_1 t}}. \tag{36}$$

The interfacial mass transfer flux of non-condensable gas

$$\dot{j}_i = -\frac{\rho_1\sqrt{D_1}(C_i - C_1)}{\sqrt{\pi t}} \tag{37}$$

is determined by the difference between the gas mass fraction in the liquid C_1 and the gas mass fraction at the interface C_i. The interface gas mass fraction, if the liquid temperature is constant, is determined as a function of pressure in following form

$$C_i = k_1 + k_2 p \tag{38}$$

where coefficients k_1 and k_2 depend on water temperature. Eq. (38) is applicable for air absorption in water and for pressures from low vacuum to several bars. According to Eq. (38), the pressure of air saturation in water is calculated as

$$p_{sat} = \frac{(C_1 - k_1)}{k_2}. \tag{39}$$

The value of the liquid pressure determines whether absorption or desorption takes place, if it is lower than the pressure of gas saturation in liquid, desorption occurs, and if it is higher, then absorption occurs.

By including Eqs. (22), (23) and (37) in Eq. (20), the interface mass transfer rate of non-condensable gas in bubbly flow can be calculated as

$$\Gamma = -\frac{6^{2/3}\rho_1 D_1^{1/2} n_b^{1/3}\alpha_2^{2/3}}{\pi^{1/3}t^{1/2}}(C_i - C_1) \tag{40}$$

and in case of annular flow, by including Eqs. (24) and (37) in Eq. (20), follows

$$\Gamma = -\frac{4\rho_1 D_1^{1/2}\alpha_2^{1/2}}{(\pi t)^{1/2} d}(C_i - C_1). \tag{41}$$

Taking into account that the change in the non-condensable gas mass fraction in the liquid is equal to the change in non-dissolved gas quality in the two-phase gas-liquid mixture, $dC_1 = -dx$ from Eq. (19) follows

$$\frac{dC_1}{dt} = -\frac{\Gamma}{\rho}. \tag{42}$$

Since the bubbly flow in air-water mixture occurs during water hammer with gaseous cavitation [6, 47], the Eq. (40) is used for the estimation of the maximum value of air mass transfer during gaseous cavitation in [48].

In technical systems with degassed water, there is a certain amount of non-condensable gas dissolved in the liquid, while a small amount of gas is undissolved and is dispersed in liquid in the form of microbubbles. When, during the transient, the rarefaction wave propagates, the volume of microbubbles increases due to the pressure drop. After the passage of the rarefaction wave, the gas microbubbles can interact with each other and they may merge. The dynamic change of the bubble radius is expressed by the Rayleigh-Plesset equation, obtained in [48],

$$\frac{p_b - p_1}{\rho_1} = r_b \frac{d^2 r_b}{dt^2} + \frac{3}{2}\left(\frac{dr_b}{dt}\right)^2 + \frac{4\mu_1}{\rho_1 r_b}\frac{dr_b}{dt} + \frac{2\sigma}{\rho_1 r_b} \tag{43}$$

where p_b is pressure in the gas bubble, p_1 is the water pressure, r_b is the bubble radius, σ is the surface tension, μ_1 is dynamic viscosity of water, and ρ_1 is density of water. A detailed numerical solution, for air bubble growth in water at room temperature under a sudden pressure drop, is presented in [48], as well as the analysis of the influence of the initial bubble radius and the pressure drop value on the bubble growth dynamics. It is concluded that the rate of mass transfer is significantly higher in the initial period of gaseous cavitation than during the rest of the transient. This period is approximately equal to the time steps of the numerical integration of governing differential equations. For instance, in the case of the MOC, the time step of integration is determined by the Courant criterion, see Eq. (55) below. The time step of integration of the governing balance equations is no smaller than 10^{-4} s, which is of the same order of magnitude as the time period of inertial bubble growth. With regard to these effects, for the first step of integration of the balance equation when the liquid pressure drops below the pressure of gas saturation in liquid, the rate of mass transfer is calculated with the following empirical equation

$$dx = -dC_1 = \frac{\Gamma}{\rho}dt = -k(C_i - C_1) \tag{44}$$

where the value of coefficient $k = 0.7735$ is determined in [48], by comparing the measured and calculated pressure changes during the transient with gaseous cavitation. After this first step of integration, it is assumed that the finite number of air bubbles is formed, which results in the bubble number density $n_b \approx 10^8$, while Eq. (40) is later applied for the mass transfer rate calculation.

The pipes wall elasticity is taken into account in determination of the speed of pressure wave propagation

$$c = a\left(1 + \frac{\rho a^2}{E}\frac{d}{\delta}\right)^{-1/2} \tag{45}$$

where E is Youngs's modulus of elasticity of the pipe, δ is the pipe wall thickness, and a is the sonic velocity. In case of the single-phase water flow,
the sonic velocity [38] is calculated as

$$a = \sqrt{\left(\frac{dp}{d\rho}\right)_s} \tag{46}$$

and in the homogeneous two-phase flow without evaporation or condensation, the sonic velocity is calculated from the so-called frozen sonic velocity expression [49]

$$a = \left[\rho\left(\frac{\alpha_2}{\rho_2 c_2^2} + \frac{1-\alpha_2}{\rho_1 c_1^2}\right)\right]^{-\frac{1}{2}} \tag{47}$$

where c_1 and c_2 are the sonic velocities in liquid and vapor, respectively, and they are calculated using Eq. (46), for known values of thermodynamic parameters of liquid and vapor.

Presented closure laws for boiling and condensation are incorporated into the hydrodynamic model presented in Section 2.1, and the results obtained with the numerical solutions show an interplay of hydrodynamic effects of pressure waves generation and propagation and interfacial mass and energy transfer by phase transition and diffusion. It is interesting to note that an interaction of the thermal and hydrodynamic effects is observed even at the nanoscale level by the molecular dynamic simulations of boiling [50].

2.4 Numerical solution

Governing equations Eqs. (1)-(3) are transformed, including application of material derivative, introducing sonic velocity in one-phase fluid, and thus in homogeneous two-fluid model, as a function of pressure and enthalpy

$$c = \left(\left(\frac{\partial\rho}{\partial p}\right)_h + \frac{1}{\rho}\left(\frac{\partial\rho}{\partial h}\right)_p\right)^{-1/2} \tag{48}$$

and grouping of all partial differentials over time and coordinate on the left-hand side of equations, so that the system of quasi-linear hyperbolic partial differential equations is obtained in following form

$$\frac{\partial p}{\partial t} + u\frac{\partial p}{\partial x} + c^2\rho\frac{\partial u}{\partial x} = X \tag{49}$$

$$\frac{\partial u}{\partial t} + u\frac{\partial u}{\partial x} + \frac{1}{\rho}\frac{\partial p}{\partial x} = Y \tag{50}$$

$$\frac{\partial h}{\partial t} + u\frac{\partial h}{\partial x} - \frac{1}{\rho}\left(\frac{\partial p}{\partial t} + u\frac{\partial p}{\partial x}\right) = Z \tag{51}$$

where

$$X = -c^2\left(\frac{\partial\rho}{\partial h}\right)_p\left(\frac{fu^2|u|}{2d_{\mathrm{H}}} + f_{\mathrm{u}}u\,\mathrm{sign}(u)\left|\frac{\partial u}{\partial t}\right| - gu\sin\theta - \int_{x_{\mathrm{i}}-\varepsilon}^{x_{\mathrm{i}}+\varepsilon} sign\left(h - h'\right)\frac{\Gamma_c h''}{\rho}dx\right) \tag{52}$$

$$Y = -\frac{fu|u|}{2d_{\mathrm{H}}} - f_{\mathrm{u}}\mathrm{sign}(u)\left|\frac{\partial u}{\partial t}\right| - g\sin\theta \tag{53}$$

$$Z = \frac{fu^2|u|}{2d_{\mathrm{H}}} + f_{\mathrm{u}}u\,\mathrm{sign}(u)\left|\frac{\partial u}{\partial t}\right| - \int_{x_i-\varepsilon}^{x_i+\varepsilon} sign\left(h - h'\right)\frac{\Gamma_c h''}{\rho}dx. \tag{54}$$

The system of equations, Eq. (49)-(51), is solved for the appropriate initial and boundary conditions by the MOC. Initial conditions are defined by thermal–hydraulic characteristics of the fluid at the initial time, before the disturbance happens. The boundary conditions are determined by the state of the fluid at the beginning and end of the sections, as well as at the boundaries of the observed system. Three characteristic paths are used, where two correspond to the pressure wave propagation (C^+ and C^-, **Figure 2**) and the third to the propagation of the fluid particle enthalpy front (C^P, **Figure 2**). The spatial step of integration Δx, that is the distance between two adjoining nodes, is constant within one pipe segment. The time step of integration is determined according to the Courant criterion:

$$\Delta t \le \min\left(\frac{\Delta x}{c_{i,j} + |u_{i,j}|}\right), \quad i = 1, 2, \ldots N(j), \; j = 1, 2, \quad \ldots \quad, M \tag{55}$$

where index i denotes the node within the pipe j.

Conservation equations Eq. (49)-(51) are transformed into the system of ordinary differential equations along three characteristic paths.

$$dp + \rho c du = (X + \rho c Y)dt \text{ along } C^+ : \frac{dt}{dx} = \frac{1}{u + c} \tag{56}$$

$$dp - \rho c du = (X - \rho c Y)dt \text{ along } C^- : \frac{dt}{dx} = \frac{1}{u - c} \tag{57}$$

$$dh - \frac{1}{\rho}dp = Zdt \text{ along } C^P : \frac{dt}{dx} = \frac{1}{u}. \tag{58}$$

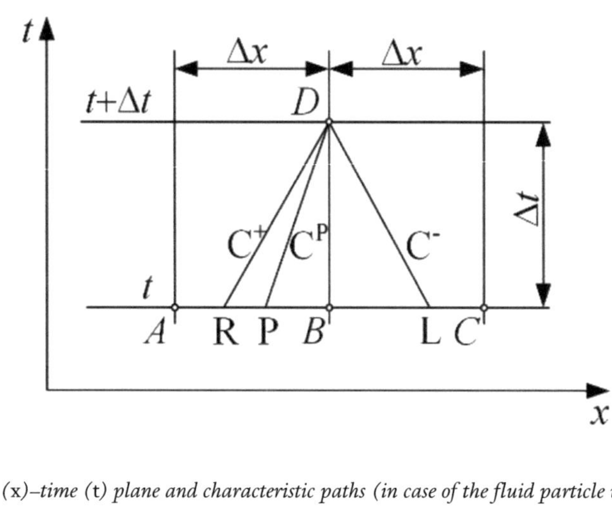

Figure 2.
Spatial coordinate (x)–time (t) plane and characteristic paths (in case of the fluid particle velocity $u_P > 0$*).*

Approximation of total derivatives by finite differences along the characteristic paths transforms differential equations, Eqs. (56)-(58), into difference equations.

A, B and C in **Figure 2** denote three successive nodes in the flow channel, which are used for calculation of initial values of the dependent variables at time level t. The pressure and velocity values in points R and L are calculated by the linear interpolation of their initial values in A, B and C. The point D denotes the intersection point of all characteristic paths C^{+}, C^{-} and C^{P} and the x-axis at time level $t + \Delta t$. Hence, D denotes the node where disturbance arrives in the next time level, $t + \Delta t$. Coordinate x_P is determined using the slope of the characteristic path C^{P} and the linear interpolation of the velocity between nodes A and B for positive flow direction and between nodes B and C for negative flow direction, **Figure 3**.

The Lagrange's interpolation polynomial (LIP) of the third degree is used for determination of the initial enthalpy value in point P with the aim of reducing the numerical diffusion of the enthalpy front propagation. For the purpose of derivation of LIP of the third degree it is necessary to use the enthalpy values at four nodes. The choice of nodes depends on the flow direction. It is always necessary to use the

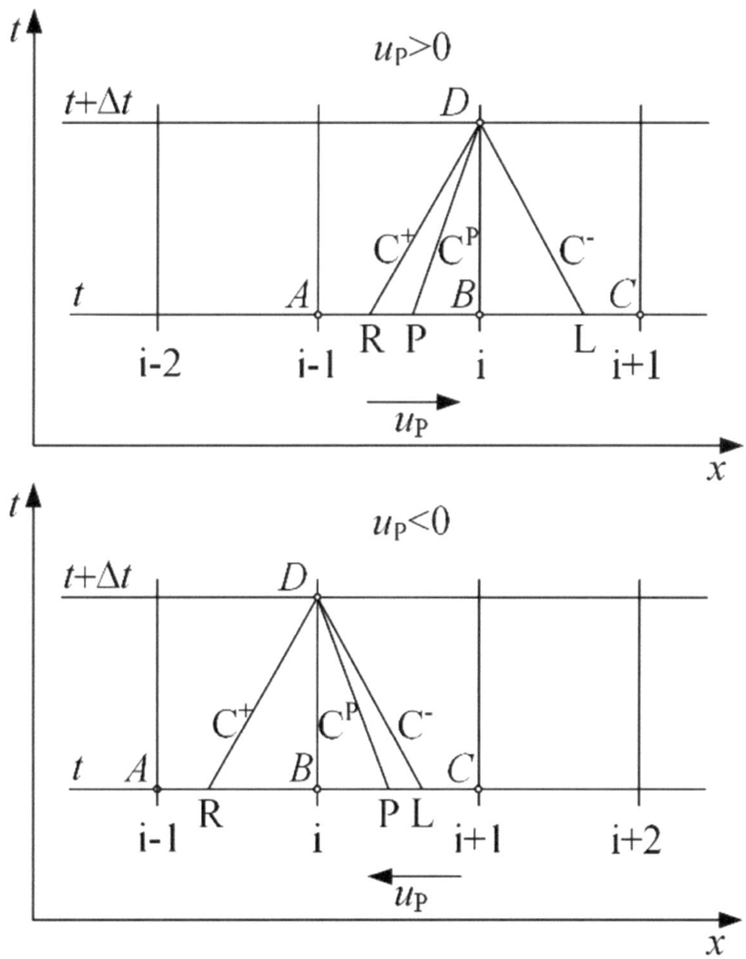

Figure 3.
Determination of nodes for the Lagrange's interpolation polynomial of the third degree, for positive ($u_P > 0$) *and negative* ($u_P < 0$) *flow direction.*

enthalpy value in one node downstream, in two nodes upstream, as well as in the observed node (designated with i in **Figure 3**). The following calculation algorithm, suitable for the computer programming of LIP, is applied:

$$L_m(x) = \Pi_{m+1}(x) \sum_{j=0}^{m} \frac{h_j}{D_j} \tag{59}$$

where

$$\Pi_{m+1}(x) = (x - x_0) \dots (x - x_m) \tag{60}$$

$$D_j = (x_j - x_0)(x_j - x_1) \dots (x_j - x_{j-1})(x - x_j)(x_j - x_{j+1}) \dots (x_j - x_m), j = 0, 1, \dots, m. \tag{61}$$

The error of the interpolation with the LIP is [51].

$$|h(x) - L_m(x)| \le \frac{M_{m+1}}{(m+1)!} |\Pi_{m+1}(x)| \tag{62}$$

and

$$M_{m+1} = \max_{a \le x \le b} \left| f^{(m+1)}(x) \right|. \tag{63}$$

The use of LIP of the third degree to determine the enthalpy at the point P gives the truncation error of the fourth order $O[(\Delta x)^4]$ for the numerical discretization of the enthalpy along the x coordinate. The integration of the energy equation Eq. (58) with respect to time is performed along the characteristic path C^P with the Euler explicit method, which gives truncation error of the first order $O(\Delta t)$.

3. Results and discussion

The developed model and computer program of transient compressible fluid flow is applied for calculation of thermal-hydraulic parameters during the CIWH event. The validation of the developed program was performed by comparing the obtained numerical results with the available results of experimental measurements from the literature [27] and comparing them with the numerical results obtained using commercial programs TUF [12] and RELAP5/MOD3 [27].

The numerical simulation of the water hammer with air cavitation caused by the rapid valve closure and the liquid column separation in case of low-velocity flow is carried out for experimental conditions of water hammer test in [6]. The saturation pressure of water is not reached, and evaporation does not occur. Measured and calculated values of pressure change at different distances from the tube entrance are compared.

The upstream type of vaporous cavitation is simulated with the developed model and presented numerical method. Calculated values of pressure head and fluid velocity, at different distances from the isolating valve, are compared with measured values in experimental installation [52].

3.1 Condensation induced water hammer in a test facility

Numerical simulation of the steam–water interface propagation and the CIWH caused by direct steam condensation was performed for conditions of a simple experimental apparatus consisting of a tank, a horizontal and a vertical pipe, of the same diameter (**Figure 4**). Zaltsgendler et al. [18] experimentally investigated CIWH in the test facility. The horizontal pipe is at one end connected to the tank. The tank and the horizontal pipe are filled with subcooled water, at initial pressure 0.551 MPa and temperature 22°C. The vertical pipe is filled with saturated steam at the pressure of 0.382 MPa. At one end, the pipe is connected by a fast-acting ball valve to the horizontal pipe, while at the other end is closed.

By opening the fast-acting ball valve at the moment t = 0 s, the subcooled water and saturated steam are brought into direct contact, and conditions are met for CIWH event. The intensive condensation of steam onto the head of the water column occurs, and the water column starts moving towards the space which was filled with steam before condensation occurred. Due to condensation, the pressure in the steam drops sharply, and the water column accelerates towards the closed end of the vertical pipe. At the moment when all the steam in the vertical tube is condensed, the head of the water column splashes the closed end of the pipe at 1.34 s and causes a pressure increase of 8.53 MPa (**Figure 5a**). Afterwards, the pressure wave propagates towards the tank, where it is reflected from the water mass, and its amplitude attenuates due to the friction on the pipe walls. This wave with attenuated amplitude moves towards the closed end of the vertical pipe and hits it, but this time with less intensity. This process is repeated periodically. The numerically obtained results are in acceptable agreement with the measured values presented in **Figure 5b**. In **Figure 5a**, it is shown that the calculated pressure pulses diminish at about 3.8 s, while the measured pressure pulses diminish after 4 s, **Figure 5b**. The cause of the enlarged difference between measured and calculated pressure pulses, after the first peak occurrence, may be in the method of determination of the evaporation and condensation rate and the sonic velocity in the two-phase mixture after the pressure wave reflection from the closed pipe end.

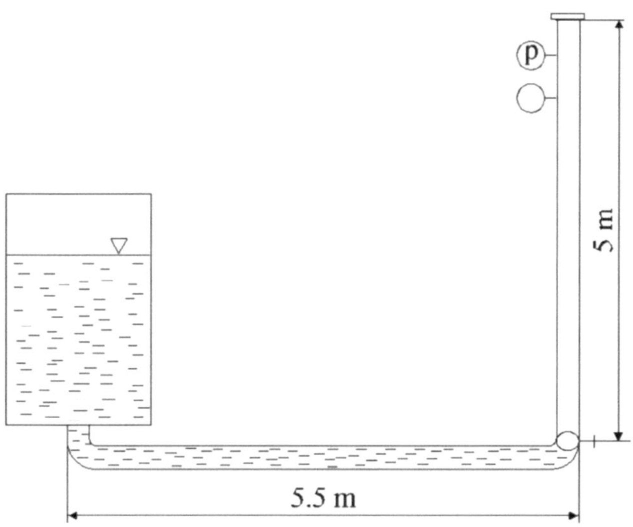

Figure 4.
Schematic view of the experimental apparatus for CIWH testing [18].

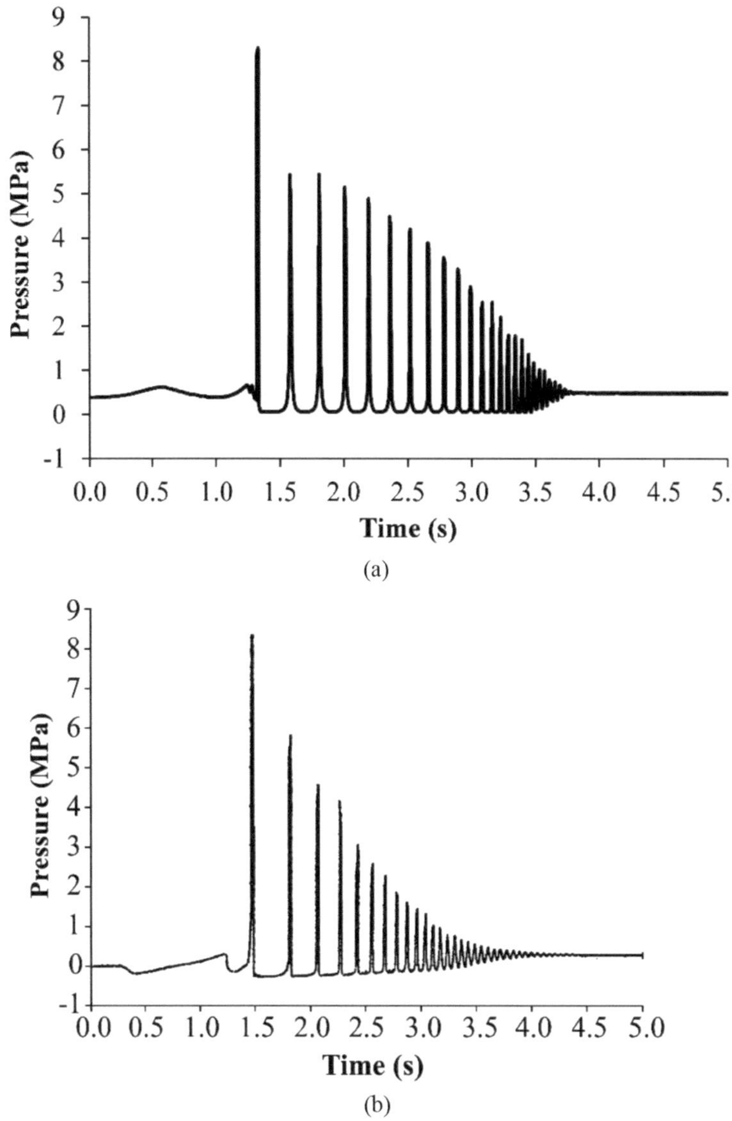

Figure 5.
Pressure change near the closed end of the vertical pipe in the experimental apparatus for CIWH testing. Results obtained by applying the developed model (up) and the results of experimental measurements [18] (down).

The first pressure pulse is the most dangerous from the point of view of equipment safety because it has the largest amplitude. By applying the developed model, a good prediction of the time of occurrence and the amplitude of the first and most dangerous pressure pulse was achieved. The presented calculation results were obtained with spatial step of 0.1 m.

3.2 Water cannon test

The CIWH in the vertical pipe for steam discharge into a pool with subcooled water, known as water cannon, was experimentally investigated in [27]. The experimental apparatus, shown in **Figure 6**, consists of a vertical metal pipe 0.7112 m long

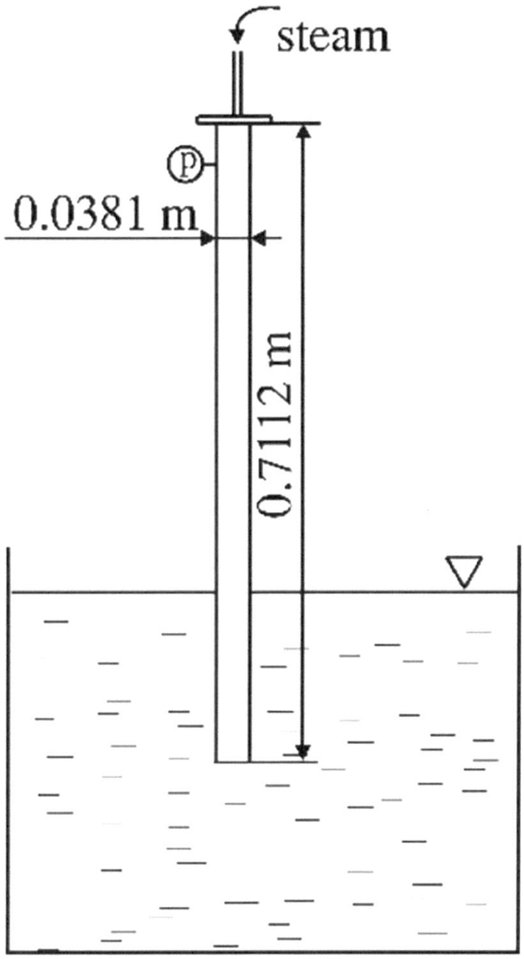

Figure 6.
Schematic view of the experimental apparatus for simulating a water cannon.

with an inner diameter of 0.0381 m and a large tank filled with subcooled water in which the pipe is immersed with its bottom to a depth of several centimeters. Pressure gauges are mounted on the top of the pipe. Saturated steam at a pressure of 0.1023 MPa is introduced into the vertical pipe at a constant speed through a small pipe, located at its top, and the lower end of the vertical pipe is immersed in a tank with water at a temperature of 49°C, at the same pressure of 0.1023 MPa. The problem is observed from the moment when the vertical pipe is completely filled with steam and the valve at its top is closed. The initial velocity in all parts of the system is equal to 0 m/s.

This situation can occur at the steam turbine exit, during the discharge of steam into a larger volume of cold water. Direct contact of steam and subcooled water leads to intensive steam condensation. The steam pressure drops, and a liquid column in the form of a plug enters the pipe. As the water level in the pipe rises, due to the difference in pressure, the condensation process is carried further towards its top, where the water column eventually hits and causes a large pressure pulse on the valve. The maximal pressure pulse occurs at the first impact.

Pressure change over time near the closed end of the vertical pipe obtained using the developed program (**Figure 7a**) is compared with pressure change obtained by the RELAP code [27] (**Figure 7b**). The intensity of the first pressure peak, which represents the risk of causing mechanical damage to the pipeline, matches the results from the literature with which they were compared. Other peaks predicted by the RELAP code (**Figure 7b**) are caused by the acoustic propagation of pressure waves in the vertical pipe between the tank and the closed end of the pipe, while the developed program (**Figure 7a**) predicts only four pulses caused by the movement of the water column in the pipe.

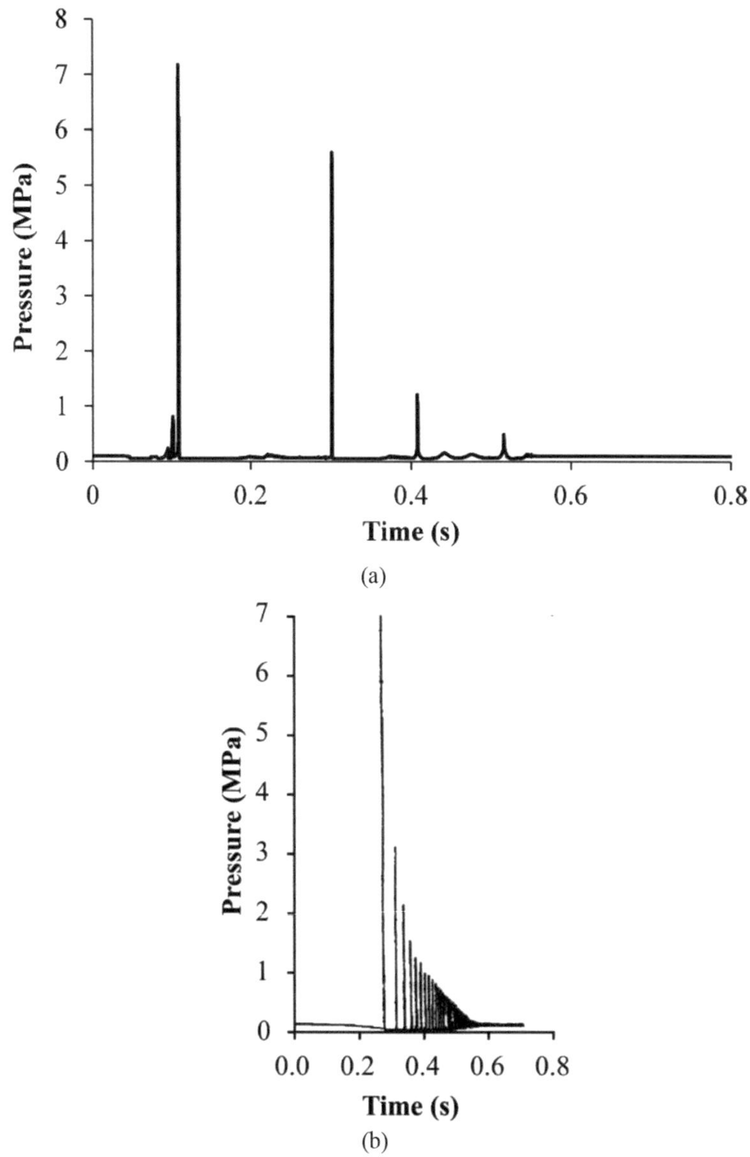

Figure 7.
Pressure change near the closed end of the vertical pipe in the water cannon test by Yeung et al. [27]. The result obtained by applying the developed model (up) and result of RELAP5/MOD3 simulation (down).

3.3 Water hammer with non-equilibrium gas release

The numerical simulation of the water hammer caused by the rapid valve closure and the liquid column separation was carried out for experimental conditions of water hammer test [6]. The experimental installation, shown in **Figure 8**, consists of two reservoirs under different pressures that are connected by a straight copper pipeline with a constant inclination to the horizontal of 3.2° and a total length of 37.23 m. The inner diameter of the tube is 0.0221 m. Demineralized water is used as the working fluid. Each performed experiment consisted of two phases. First, a stationary flow in the tube is established, the initial conditions are determined, and then, a transient is induced by rapidly closing the valve. The water hammer test is performed for the case where the fast-closing valve is upstream and water flows upwards. The initial velocity in the installation is equal to 0.3 m/s (i.e., low-velocity flow). The pressure in the reservoir 2 is 0.32 MPa, and the assumed initial water temperature is 20°C. The fast-closing valve with spring, located in front of the reservoir 1, was closed in 0.009 s in the experiment. Since the effective flow reduction in the experiment is 0.004 s, the reduction of the flow velocity in the calculation starts at 0.005 s, and at 0.009 s, the value of the velocity drops to zero. The initial value of the volume fraction of the gas phase (i.e., air micro-bubbles) is 10^{-7}, which is the value assumed in [6], and the measured sonic velocity before closing the valve is 1319 m/s. The pressure is measured at four locations: in front of the fast-closing valve (p_{v1} in **Figure 8**), at three quarters of the pipe length from the Reservoir 2 (p_{q1} in **Figure 8**), at the half-length of the tube (p_{mp} in **Figure 8**) and at one quarter of the pipe length from the reservoir 2 (p_{q2} in **Figure 8**).

The presented case of low-velocity flow leads to water hammer with air cavitation, where the saturation pressure of water has not been reached, and the evaporation does not occur. Measured and calculated values of pressure change at different distances from the tube entrance are shown in **Figure 9**. The calculated pressure values are in very good agreement with the measured values in both amplitudes and oscillation periods during the entire duration of the recorded transient process of 1.5 s. Both the measured and calculated data show a sudden pressure jump due to the rapid closing of

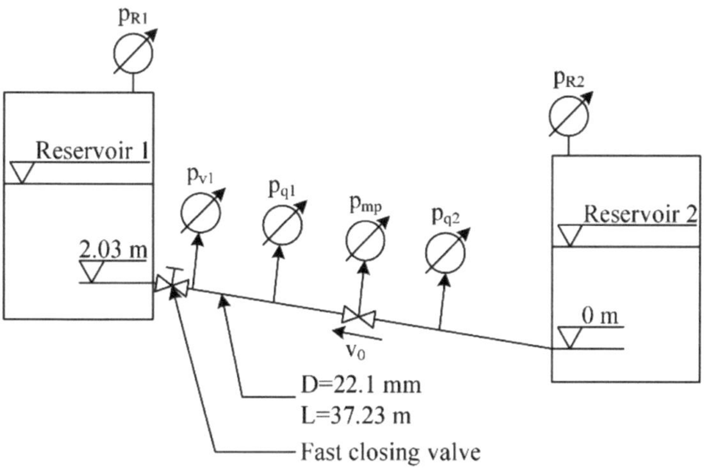

Figure 8.
Schematic view of the experimental installation for water hammer with air cavitation [6].

the valve at the beginning of the transient (**Figure 9a**). The amplitude of the pressure increase is about 0.44 MPa, which corresponds to the value determined according to the Joukowsky equation. The compression wave generated in front of the closed valve propagates towards the reservoir 2 and is reflected as rarefaction wave that travels back towards the closed valve. At 0.066 s, the rarefaction wave is reflected at the closed valve as a wave of the same sign, and the pressure drops to a low value of 0.006 MPa, which is lower than the saturation pressure of air in water, but still higher than the saturation pressure of water and steam, which is 0.0023 MPa at 20°C. The liquid column separation occurs when the gap in the water flow appears due to the presence of air, resulting from gaseous cavitation, when pressure drops below the

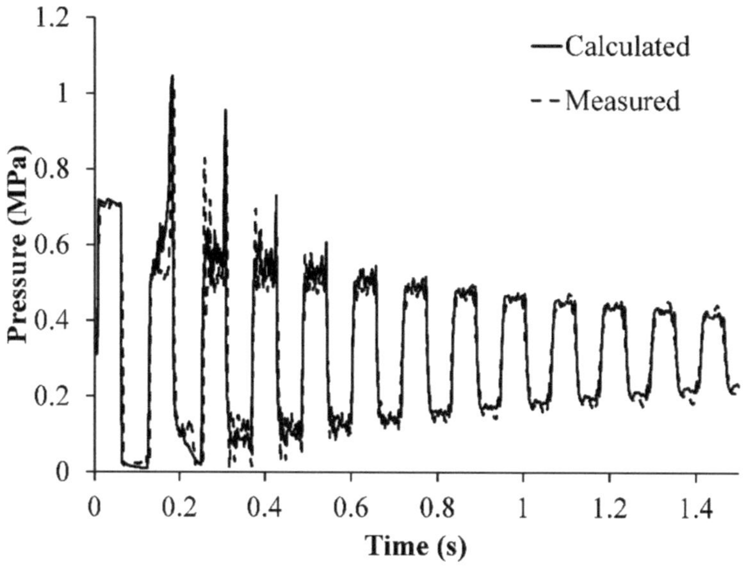

(a)

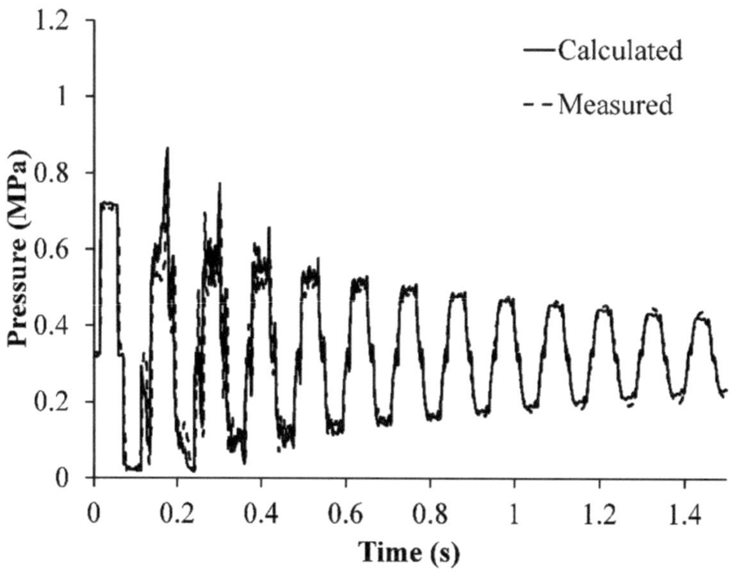

(b)

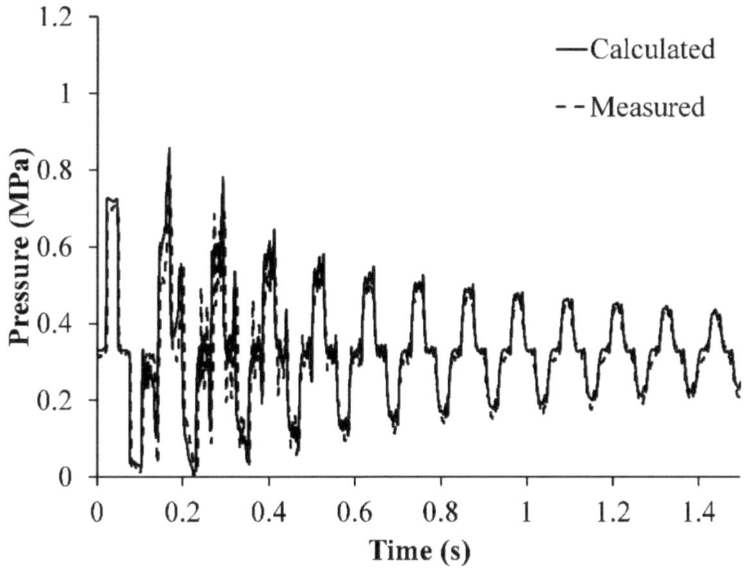

(c)

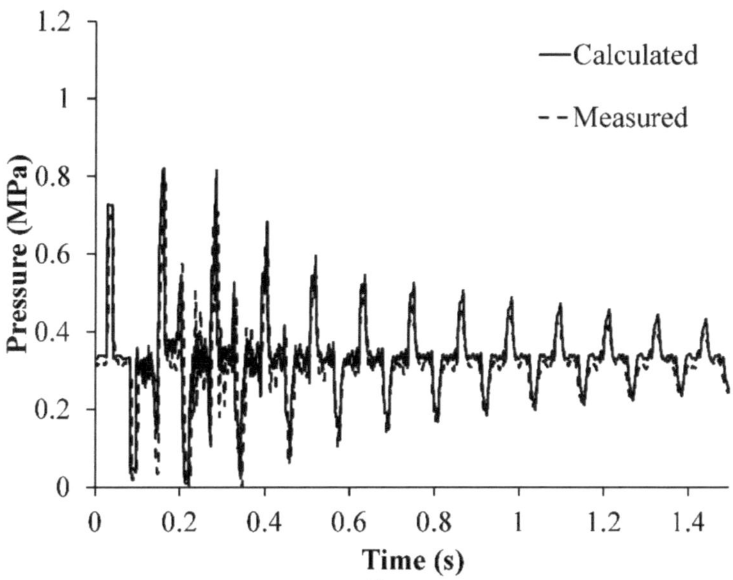

(d)

Figure 9.
Comparison of measured and calculated pressure: (a) in front of the closing valve, (b) at three quarters of the pipe length from the inlet, (c) at half of the pipe length, (d) at one quarter of the pipe length from the inlet.

saturation pressure of air in water. Since the conditions for water evaporation have not been met, there is no vaporous cavitation.

Here, presented results are obtained with a numerical discretization of the tube with 300 nodes. In the presented calculation, the initial mass fraction of dissolved air in water is estimated at 8×10^{-6}, which corresponds to the saturation of air in water at a pressure of 0.035 MPa and a temperature of 20°C. This value was chosen so as to obtain a good agreement between the calculated peak pressures and the measured data. The appropriate prediction of the friction effect is important for a more precise determination of the pressure change. It is not enough to take into account only steady friction [48].

3.4 Water hammer with vaporous cavitation

Sanada et al. [52] have experimentally investigated water hammer with vaporous cavitation in the experimental installation shown in **Figure 10**, which consists of the horizontal acrylic pipeline of 200 m length with the inner diameter of 0.0152 m connected with two reservoirs. The pipeline is filled with water. The initial water parameters are given in [30]. The focus here will be on the upstream type of vaporous cavitation.

The fast closure of upstream valve causes the pressure drop behind the valve. The pressure drops to and below the saturation pressure, the pressure at which the intensive evaporation occurs. After certain time, reflected pressure wave leads to collapse of vapor bubbles, formed in the evaporation process, and it causes pressure surge at the valve. Vapor bubbles form and disappear in cycles until the minimal steady pressure, higher than the saturation pressure, sets at the valve. This is column separation which occurs downstream, behind the obstacle which stops fluid flow.

The upstream vaporous cavitation is simulated with the model and numerical solution method presented in Section 2. Calculated values of absolute pressure head in meters and fluid velocity, at different distances from the isolating valve, are compared with measured values in experimental installation [52] in **Figures 11** and **12**. **Figure 13** shows the comparison between measured pressure values in experimental installation [52] and numerical predictions from the various models for upstream cavitation obtained by [30]. For this severe case of vaporous cavitation, all predictions of Shu show similar characteristics. The best agreement is shown in case of frequency-dependent friction model, **Figure 13d**. Comparison of measured and calculated values of the pressure change in **Figure 11** and of the change in velocity in **Figure 12** shows satisfactory agreement for the amplitudes of the pressure wave and the periods of their oscillation. Compared to the results by Shu in [30] using a DVCM, a homogeneous equilibrium model and an improved transient friction model (**Figure 13**), the developed model provides a better prediction of the attenuation of the pressure wave amplitudes in period after 6 s.

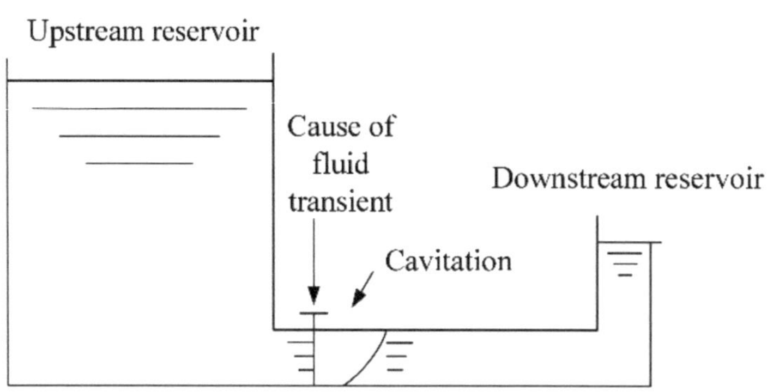

Figure 10.
Schematic view of the experimental installation for column separation investigation—upstream type [52].

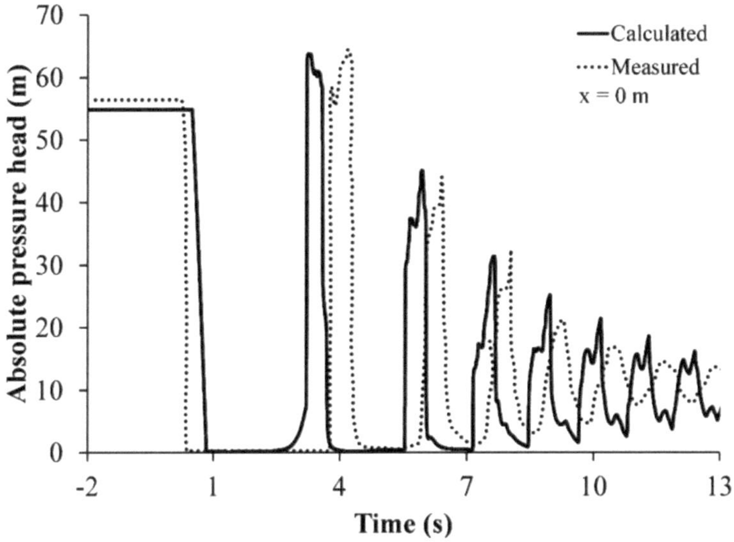
Calculated
Measured
x = 0 m
Absolute pressure head (m)
70
60
50
40
30
20
10
0
-2
1
4
7
10
13
Time (s)

(a)

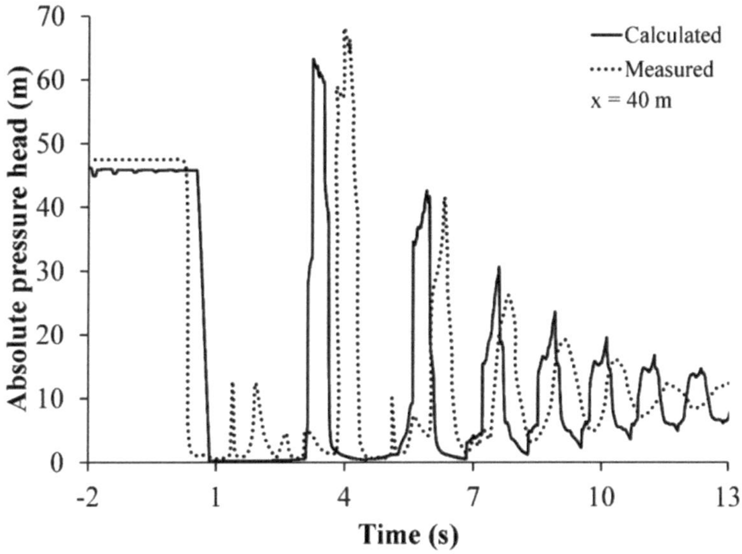
Calculated
Measured
x = 40 m
Absolute pressure head (m)
70
60
50
40
30
20
10
0
-2
1
4
7
10
13
Time (s)

(b)

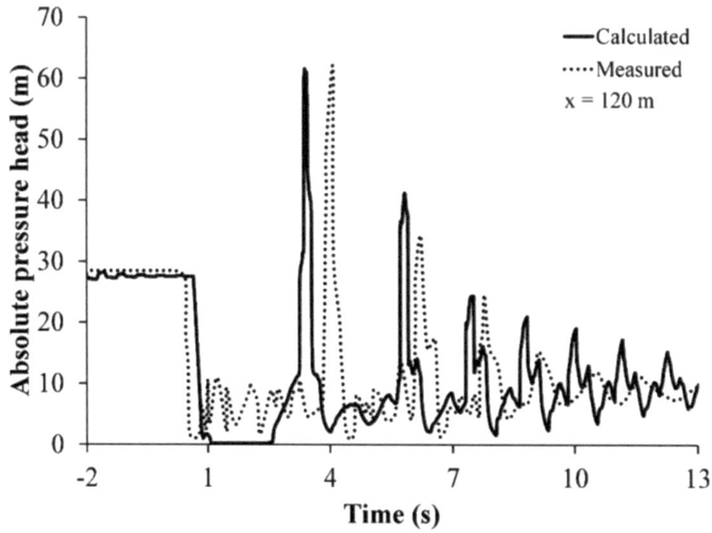

(c)

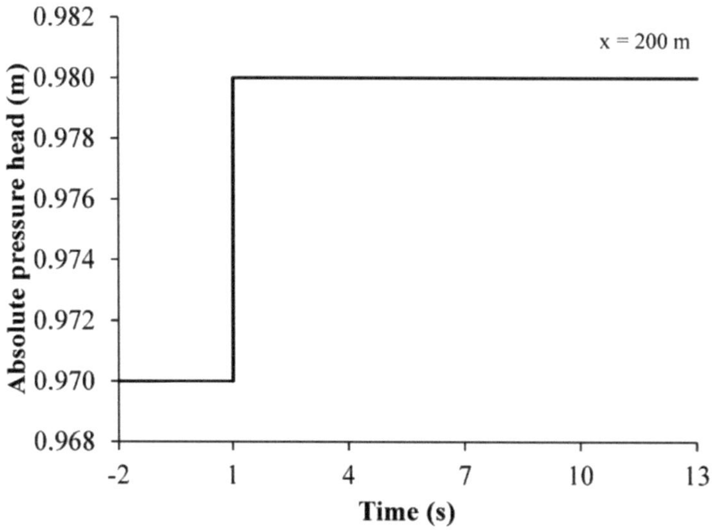

(d)

Figure 11.

Comparison of measured and calculated absolute pressure head during column separation at different distances from the isolating valve: (a) x = 0 m, (b) x = 40 m and (c) x = 120 m. (d) Calculated absolute pressure head during column separation at the distance x = 200 m from the isolating valve.

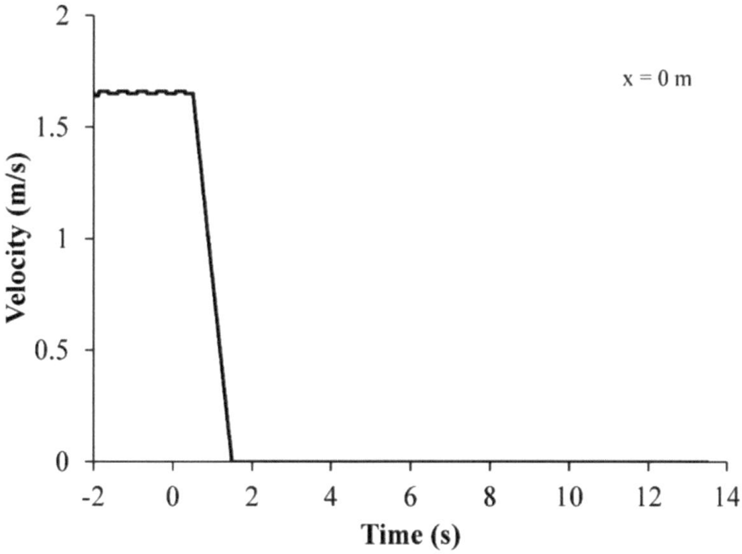
x = 0 m
Velocity (m/s)
Time (s)
2
1.5
1
0.5
0
-2
0
2
4
6
8
10
12
14

(a)

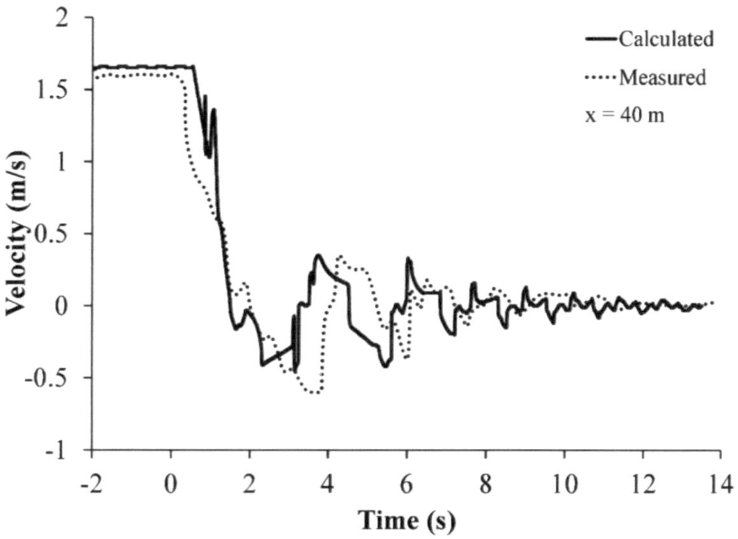
Calculated
Measured
x = 40 m
Velocity (m/s)
Time (s)
2
1.5
1
0.5
0
-0.5
-1
-2
0
2
4
6
8
10
12
14

(b)

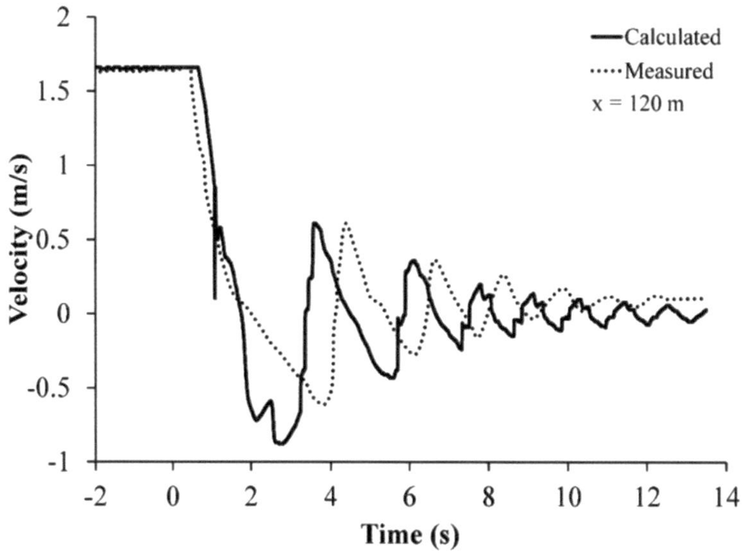

(c)

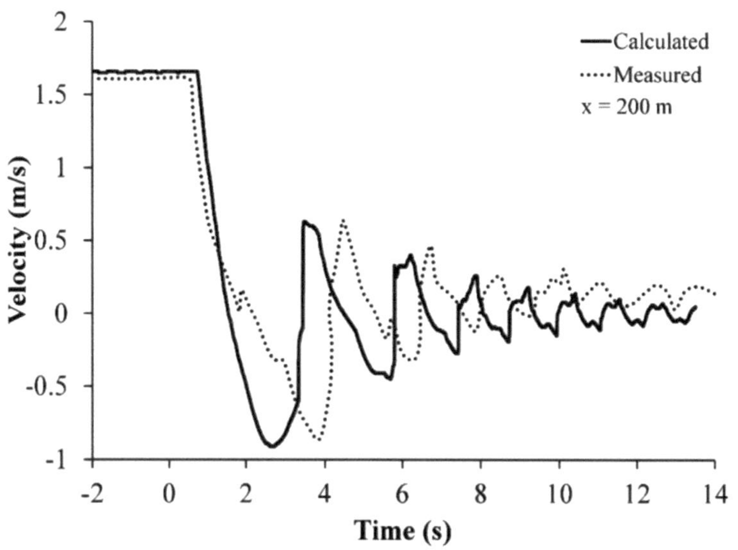

(d)

Figure 12.
(a) Calculated fluid velocity during column separation at the location of the isolating valve. Comparison of measured and calculated fluid velocity during column separation at different distances from the isolating valve: (b) x = 40 m, (c) x = 120 m and (d) x = 200 m.

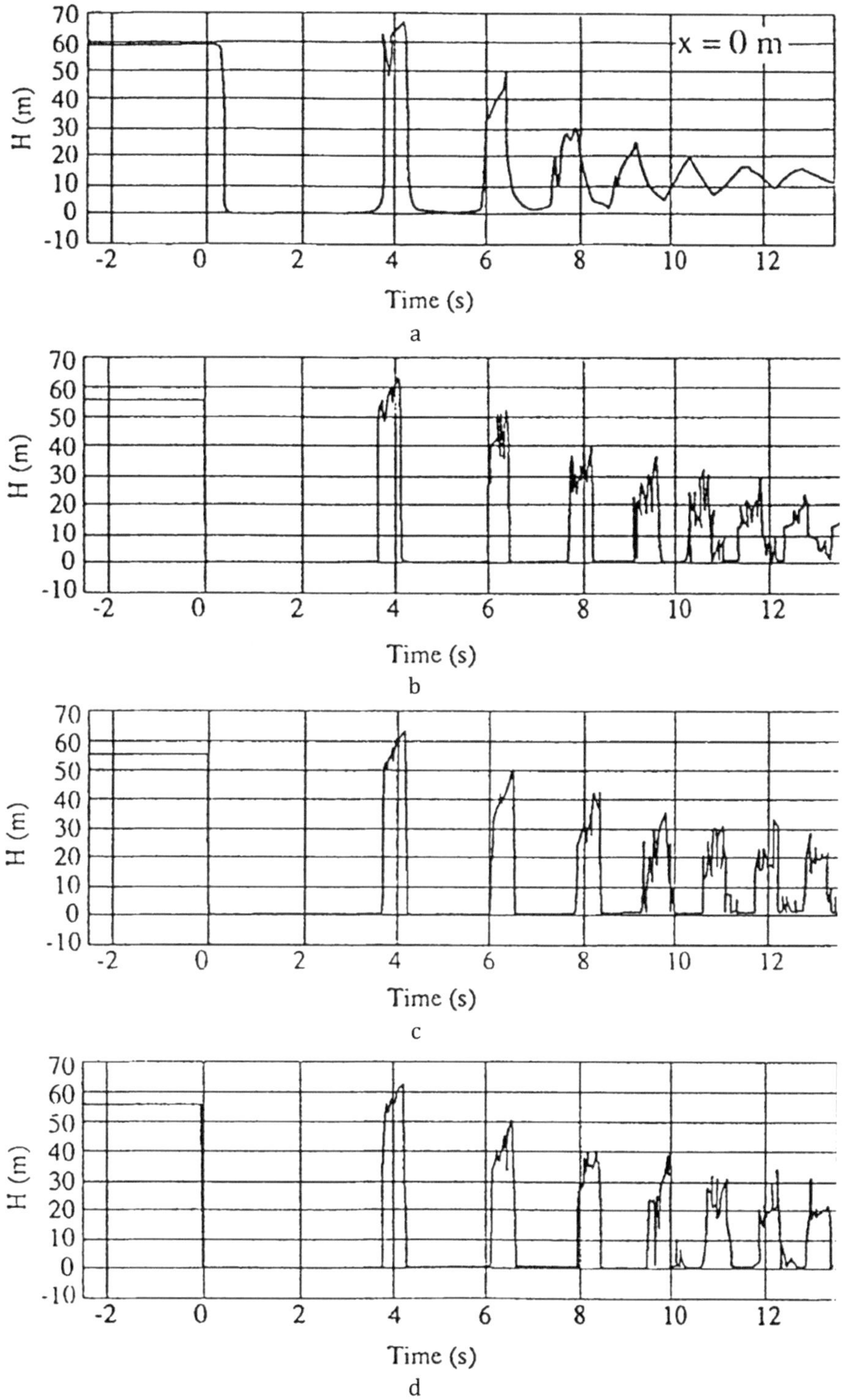

Figure 13.
Upstream vaporous cavitation. The absolute pressure changes at the isolating valve: (a) experimental results [52], (b) column separation model, (c) two-phase homogeneous equilibrium vaporous cavitation model and (d) frequency-dependent friction model [30].

4. Conclusions

The one-fluid thermal-hydraulic model is developed. The same governing equations are used for simulation and analysis of CIWH and water hammer with gaseous and vaporous cavitation. The model is based on one-dimensional, transient and compressible fluid flow of homogeneous fluid. Closure laws are developed with certain specificities for both phenomena. The model is solved by the application of MOC.

In the case of CIWH simulation, the energy equation is solved with the application of the LIP of the third degree for the calculation of initial enthalpy values at the characteristic paths of the fluid particles propagation. The prediction of the pressure pulses in CIWH events is validated by comparing with the available experimental measurements. The importance of the liquid column head tracking and the transient friction calculation for CIWH prediction is demonstrated. Previous experimental observations showed that the scatter of the test data with the same test conditions exists. This scattering is obtained also by numerical simulations in this study, and it is taken into account by modeling of the product of the condensation heat transfer coefficient and the interfacial area concentration which are correlated with the acceleration of the liquid column head and vapor interface. From the point of view of plant safety, the most important outcome of the numerical simulation of CIWH is the prediction of the first pressure pulse caused by the liquid column splashing.

A new model for water hammer with gaseous and vaporous cavitation is presented. The difference in modeling using the presented model compared to standard models is reflected in the fact that in this model the two-phase mixture exists anywhere along the pipe length, and consequently, the speed of wave propagation is equal to the sonic velocity in the two-phase mixture. Also, the model has no restrictions on the minimum pressure value during the transient, which is determined by the dynamics of the propagation of the pressure waves and the intensity of rarefaction waves. The new modeling approach is validated by comparing the obtained numerical results with the measurement results of an experimental test of the water hammer caused by the rapid valve closure on the experimental installation [6]. The present model is applicable to water hammer with gaseous and vaporous cavitation. The closure laws consist of the non-equilibrium model of gas release and absorption and desorption in water in case of non-condensable gaseous cavitation, as well as evaporation and condensation model in cases of vaporous cavitation. This modeling approach is sensitive to the spatial integration step, as a consequence of the large nonlinearity of the change of two-phase flow parameter, like the speed of pressure wave propagation and the two-phase mixture density, with changes of gas volume fraction. One of the advantages of this method is a simpler algorithm, than in standard methods, that can be easily implemented in computer programs. The gas release rate has uppermost value during the propagation of the first rarefaction wave, which causes the pressure to drop below the value of the saturation pressure of the gas in the water. In the sequel of the transient, when the gas is already released and the gas bubbles or pockets are formed, the rates of gas degassing or absorption are of a smaller order of magnitude. The difference in the gas generation rate could be the consequence of disturbed conditions in water, with dissolved gas which undergoes sudden pressure drop caused by rarefaction wave propagation during the transient. The developed model of gas release during water hammer with gaseous cavitation is justified by the comparison of the obtained numerical results with the experimental data for one research case.

Acknowledgements

The authors are grateful to Dr. Anton Bergant for providing the experimentally measured data.

This research was supported by the Ministry of Education, Science and Technological Development of the Republic of Serbia (grant 451-03-68/2022-14/200105 and grant 451-03-68/2022-14/200213 of 4.2.2022.).

Nomenclature

A	area, m^2
a	interfacial area concentration, 1/m
a	sonic velocity, m/s
a	thermal diffusivity, m^2/s
C	constant in Eq. (10)
C	mass fraction
c	speed of pressure wave propagation, m/s
D	diffusion coefficient, m^2/s
d	inside diameter, m
E	Young's modulus of elasticity, N/m^2
f	friction coefficient
g	gravity, m/s^2
H	pressure head, m
H_c	Henry's constant, Pa
h	specific enthalpy, J/kg
h	heat transfer coefficient, W/m^2 K
j	mass flux, kg/m^2 s
k	coefficient of gas interface mass fraction in Eq. (38), k_1 and k_2, 1/Pa
k	coefficient of mass transfer in Eq. (44)
L	characteristic length, m
n	concentration, $1/m^3$
p	pressure, Pa
q	surface heat flux, W/m^2
R_g	gas constant, J/kg K
r	latent heat of evaporation/condensation, J/kg
r	radius, m
T	temperature, K
t	time, s
t	renewal time in Eqs. (40) and (41), s
u	velocity, m/s
v	specific volume, m^3/kg
x	spatial coordinate, m
x	mass fraction
x	quality
$\tilde{x}$	molar fraction in Eq. (32)

Greek letters

α	void fraction
Γ	mass transfer rate, kg/m^3 s
δ	thickness, m
ε	infinitesimal distance, m
θ	angle of inclination, °
λ	thermal conductivity, W/m K
μ	dynamic viscosity, Pa s
ν	kinematic viscosity, m^2/s
ρ	density, kg/m^3
σ	surface tension, N/m

Non-dimensional numbers

$\mathrm{Nu} = \frac{hL}{\lambda}$	Nusselt number
$\mathrm{Pr} = \frac{\nu}{a}$	Prandtl number
$\mathrm{Re} = \frac{ud}{\nu}$	Reynolds number
$\mathrm{We} = \frac{\rho_1 (u_2 - u_1)^2 d_b}{\sigma}$	Weber number

Subscripts or superscripts

$'$	saturated liquid state
$''$	saturated vapor state
0	initial condition
1	liquid
2	gas/vapor
b	bubble
c	condensation
D	droplet
ED	Entrained Droplet
eq	equilibrium condition
H	hydraulic
h	isenthalpic process
i	interface
LCH	Liquid Column Head
p	isobaric process
s	isentropic process
sat	saturation
t	thermodynamic
u	unsteady

Acronyms and abbreviations

CFD	Computational Fluid Dynamics
CIWH	Condensation Induced Water Hammer

DGCM	Discrete Gas Cavitation Model
DVCM	Discrete Vapor Cavitation Model
ED	Entrained Droplets
FIIWH	Flashing Instability Induced Water Hammer
FSI	Fluid Structure Interaction
GIVCM	Generalized Interface Vapor Cavitation Model
LCH	Liquid Column Head
LIP	Lagrange's Interpolation Polynomial
MOC	Method of Characteristics
NCS	Natural Circulation System
NPP	Nuclear Power Plant
ROSA	Rig-Of-Safety Assessment (ROSA)
SCIF	Steam Condensation Induced Flashing

Author details

Sanja Milivojevic[1]*, Vladimir Stevanovic[1], Milan M. Petrovic[1] and Milica Ilic[2]

1 Faculty of Mechanical Engineering, University of Belgrade, Belgrade, Serbia

2 Innovation Centre of the Faculty of Mechanical Engineering, University of Belgrade, Belgrade, Serbia

*Address all correspondence to: smilivojevic@mas.bg.ac.rs

References

[1] Wang L, Yue X, Chong D, Chen W, Yan J. Experimental investigation on the phenomenon of steam condensation induced water hammer in a horizontal pipe. Experimental Thermal and Fluid Science. 2018;**91**:451-458. DOI: 10.1016/j.expthermflusci.2017.10.036

[2] Sun J, Lu C, Mi Z, Cao X, Ding M. Experimental research on characteristics of condensation induced water hammer in natural circulation systems. International Communications in Heat and Mass Transfer. 2020;**114**:104559. DOI: 10.1016/j.icheatmasstransfer.2020.104559

[3] Hammitt FG. Cavitation and Multiphase Flow Phenomena. New York, USA: McGraw-Hill Inc.; 1980. p. 423. ISBN: 0070259070

[4] Kolev NI. Multiphase Flow Dynamics 3: Turbulence, Gas Absorption and Release, Diesel Fuel Properties. Berlin, Germany: Springer; 2007. p. 308. DOI: 10.1007/978-3-540-71443-9

[5] Ilic MM, Petrovic MM, Stevanovic VD. Boiling heat transfer modelling—A review and future prospectus. Thermal Science. 2019; **23**(1):87-107. DOI: 10.2298/TSCI180725249

[6] Bergant A, Simpson AR. Pipeline column separation flow regimes. Journal of Hydraulic Engineering. 1999;**125**:835-848. DOI: 10.1061/(ASCE)0733-9429(1999)125:8(835)

[7] Bergant A, Simpson AR, Tijsseling AS. Water hammer with column separation: A historical review. Journal of Fluids and Structures. 2006;**22**:135-171. DOI: 10.1016/j.jfluidstructs.2005.08.008

[8] Miwa T, Sano M, Yamamoto K. Experimental studies on water hammer phenomenon including vapor column separation. Water Supply. 1990;**8**: 430-438

[9] Proovost GA, Wylie EB. Discrete Gas Model to Represent Distributed Free Gas in Liquids. In: Proceedings of the Fifth International Symposium on Water Column Separation – IAHR; 28-30 September 1981; Obernach, West Germany: IAHR; 1981. p. 8

[10] Milivojevic S, Stevanovic V, Maslovaric B. Condensation induced water hammer: Numerical prediction. Journal of Fluids and Structures. 2014; **50**:416-436. DOI: 10.1016/j.jfluidstructs.2014.07.003

[11] de Vries M, Simon A. Suction Effects on Feedpump Performance: A Literature Survey. Palo Alto, USA: Sulzer Brothers Limited, EPRI Report CS-4204; 1985

[12] Liu WS, Tahir A, Zaltsgendler E, Kelly W, Leung RK. Development Status of TUF Code. In: Proceedings of the 17th Annual Canadian Nuclear Society Conference; 9–12 June 1996; Fredericton, Canada: CNS; 1996. p. 17

[13] Serkiz AW. An Evaluation of Water Hammer in Nuclear Power Plants. In: Proceedings of the 2nd International Topical Meeting on Nuclear Reactor Thermalhydraulics; 11-13 January 1983; Santa Barbara, California, USA: NURETH; 1983. pp. 803-807

[14] Beuthe TG. Review of Two-Phase Water Hammer. In: Proceedings of the 18th Annual Canadian Nuclear Society Conference; 8-11 June 1997; Toronto, Canada: CNS; 1997. p. 20

[15] Kirsner W. Waterhammer. HPAC Heating/Piping/Air/Conditioning. 1999; **71**(2):112-122

[16] Martin CS. Condensation-induced water hammer in a horizontal pipe. Kerntechnik. 2012;77(2):94-100. DOI: 10.3139/124.110237

[17] Gruel RL, Huber PW, Hurwitz WM. Piping response to steam-generated water hammer. Journal of Pressure Vessel Technology. 1981;**103**(3):219-225. DOI: 10.1115/1.3263394

[18] Zaltsgendler E, Tahir A, Leung RK. Condensation-induced waterhammer in a vertical upfill pipe. Transactions of the American Nuclear Society. 1996;**74**:346

[19] Sun J, Deng J, Ran X, Cao X, Fan G, Ding M. Experimental research on the mechanisms of condensation induced water hammer in a natural circulation system. Nuclear Engineering and Technology. 2021;**53**:3635-3642. DOI: 10.1016/j.net.2021.05.035

[20] Hibiki T, Rassame S, Liu W, Wang L, Zhao Q, Chong D, et al. Modeling and simulation of onset of condensation-induced water hammer. Progress in Nuclear Energy. 2020;**130**: 103555. DOI: 10.1016/j.pnucene.2020. 103555

[21] Pham TQD, Choi S. Numerical analysis of direct contact condensation-induced water hammering effect using OpenFoam in realistic steam pipes. International Journal of Heat and Mass Transfer. 2021;**171**:121099. DOI: 10.1016/ j.ijheatmasstransfer.2021.121099

[22] Urban C, Schlüter M. Investigations on the stochastic nature of condensation induced water hammer. International Journal of Multiphase Flow. 2014;**67**:1-9. DOI: 10.1016/j.ijmultiphaseflow.2014. 08.001

[23] Bjorge RW, Griffith P. Initiation of water hammer in horizontal and nearly horizontal pipes containing steam and subcooled water. Journal of Heat Transfer. 1984;**106**(4):835-840. DOI: 10.1115/1.3246760

[24] Chun MH, Yu SO. A parametric study and a guide chart to avoid condensation/induced water hammer in a horizontal pipe. Nuclear Engineering and Design. 2000;**201**(2):239-257. DOI: 10.1016/S0029-5493(00)00280-6

[25] Barna IF, Ezsol G. Multiple condensation induced water hammer events, experiments and theoretical investigations. Kerntechnik. 2011;**76**: 231-236. DOI: 10.3139/124.110154

[26] Hou X, Sun Z, Su J, Fan G. An investigation on flashing instability induced water hammer in an open natural circulation system. Progress in Nuclear Energy. 2016;**93**:418-430. DOI: 10.1016/j.pnucene.2016.09.015

[27] Yeung WS, Wu J, Fernandez RT, Sundaram RK. RELAP5/MOD3 simulation of the water cannon phenomenon. Nuclear Technology. 1993; **101**:244-251. DOI: 10.13182/NT93-A34786

[28] Barna IF, Inre AR, Baranyai G, Ezsol G. Experimental and theoretical study of steam condensation induced water hammer phenomena. Nuclear Engineering and Design. 2010;**240**(1): 146-150. DOI: 10.1016/j.nucengdes. 2009.09.027

[29] Seo JS, Bankoff SG. Entrainment and condensation effects in the upward acceleration of a liquid column. International Journal of Multiphase Flow. 1989;**15**(6):925-935

[30] Shu J. Modelling of vaporous cavitation on fluid transients.

International Journal of Pressure Vessels and Piping. 2003;**80**(3):187-195. DOI: 10.1016/S0308-0161(03)00025-5

[31] Vardy AE, Brown JMB. Transient turbulent friction in smooth pipe flows. Journal of Sound and Vibration. 2003; **259**(5):1011-1036. DOI: 10.1006/jsvi.2002.5160

[32] Shimada M, Brown J, Vardy A. Estimating friction errors in MOC analyses of unsteady pipe flows. Computers & Fluids. 2007;**36**(7): 1235-1246. DOI: 10.1016/j.compfluid.2006.11.005

[33] Kucienska B, Seynhaeve JM, Giot M. Friction relaxation model for fast transient flows application to water hammer in two-phase flow: The WAHA code. International Journal of Multiphase Flow. 2008;**34**(2):188-205. DOI: 10.1016/j.ijmultiphaseflow.2007.10.001

[34] Neuhaus T, Dudlik A. Experiments and comparing calculations on thermohydraulic pressure surges in pipes. In: Proceedings of the 11th International Topical Meeting on Nuclear Reactor Thermal—Hydraulics (NURETH11); 2-6 October 2005; Avignon, France. NURETH; 2005. NURETH11-540, 19 p

[35] Bloemeling F, Neuhaus T, Schaffrath A. Development and validation of DYVRO for the simulation of condensation induced water hammer: comparison with a two-phase slug model and with experimental data of the water cannon experiment. In: Proceedings of the 15th International Topical Meeting on Nuclear Reactor Thermal – Hydraulics (NURETH15); 12-17 May 2013; Pisa, Italy. NURETH; 2013. NURETH15 – 307, 14 p

[36] Kronig R, Brink JC. On the theory of extraction from falling droplets. Applied Science Research. 1951;**2**:142-154. DOI: 10.1007/BF00411978

[37] Kuznetsov YN. Heat Transfer in Safety Problems of Nuclear Reactors. Moscow: Energoatomizdat; 1989. p. 296

[38] Ghiaasiaan SM. Two-Phase Flow, Boiling, and Condensation. Cambridge, UK: Cambridge University Press; 2008. p. 613. ISBN: 978-0-521-882

[39] Saito T, Hughes DD, Carbon MW. Multi-fluid modeling of annular two-phase flow. Nuclear Engineering and Design. 1978;**50**:225-271. DOI: 10.1016/0029-5493(78)90041-9

[40] Wagner W, Kretzschmar HJ. International Steam Tables. Berlin: Springer-Verlag; 2007. p. 390. DOI: 10.1007/978-3-540-74234-0

[41] Kroeger PG. Application of a Non-Equilibrium Drift Flux Model to Two-Phase Blowdown Experiments. In: OECD/NEA Specialists Meeting on Transient Two-phase Flow; 3 August 1976; Toronto, Canada: OECD/NEA; 1976. p. 60

[42] Bauer EG, Houdayer GR, Sureau HM. A Non-Equilibrium Axial Flow Model and Application to Loss-Of-Accident Analysis. In: Proceedings of the CSNI Specialists Meeting; 3-4 August 1976; Toronto, Canada: Atomic Energy of Canada; 1976. Vol. 1, p. 429-457

[43] Zapolski PD, Bilicki Z, Bolle L, Franco J. The non-equilibrium relaxation model for one-dimensional flashing liquid flow. International Journal of Multiphase Flow. 1996;**22**:473-483. DOI: 10.1016/0301-9322(95)00078-X

[44] Blinkov VN, Frolov SD. Modeling of flashing flow in nozzles. Engineering and Physical Journal. 1982; **42**:741-746

[45] Riznic J, Kojasoy G, Zuber N. On the spherically phase change problem. International Journal of Fluid Mechanics Research. 1999;**26**(2):110-145. DOI: 10.1615/InterJFluidMechRes.v26.i2.10

[46] Nakoryakov VE, Grigoryeva NI. Nonisothermal absorption in thermotransformers. Journal of Engineering Thermophysics. 2010;**19**: 196-271. DOI: 10.1134/S1810232810040028

[47] Zielke W, Perko H-D, Keller A. Gas Release in Transient Pipe Flow. In: Proceedings of the Sixth International Conference on Pressure Surges; 4-6 October 1989; Cambridge, England: BHRA, Cranfield; 1989. p. 3-13

[48] Ivljanin B, Stevanovic V, Gajic A. Water hammer with non-equilibrium gas release. International Journal of Pressure Vessels and Piping. 2018;**165**: 229-240. DOI: 10.1016/j.ijpvp.2018.07.009

[49] Grolmes MA, Fauske HK. Comparison of the propagation of compression and rarefaction pressure pulses in two-phase, one-component bubble flow. Transactions of the American Nuclear Society. 1968;**11**(2): 683-688

[50] Ilic M, Stevanovic V, Milivojevic S, Petrovic MM. New insights into physics of explosive water boiling derived from molecular dynamics simulations. International Journal of Heat and Mass Transfer. 2021;**172**:article 121141. DOI: 10.1016/j.ijheatmasstransfer.2021.121141

[51] Demidovitch B, Maron I. Elements de calcul numerique. Moscow: Editions Mir; 1987. p. 677

[52] Sanada K, Kitagawa A, Takenaka T. A study on analytical methods by classification of column separations in a water pipeline. Transactions of the Japan Society of Mechanical Engineers Series B. 1990;**56**(523):585-593. DOI: 10.1299/kikaib.56.585

Chapter 5

Study on Heat Transfer Mechanism of Steam Condensation on Water Jet in Steam Injector

Yasuo Koizumi

Abstract

In this chapter, the heat transfer characteristics in the steam injector that has been proposed to introduce into boiling water reactors as a feed water heat exchanger and a safety injection pump are examined. The temperature and the velocity distribution in the injector were measured. The heat transfer rate from the steam flow around the water jet to the water jet was greatly larger than that of the usual turbulent flow in a pipe. High-speed camera pictures revealed the surface of the water jet was very wavy. It was supposed that the wavy motion on the water jet surface created the effective large-internal circulation flow in the water jet, which resulted in the tremendously effective heat transport from the surface into the center portion of the water jet. From the high-speed camera pictures, the characteristics of waves on the surface; the wave height, the wave velocity, and the wave length were obtained. In addition, the dimensionless numbers were found from the parameters that related to the phenomena in the steam injector. By using these dimensionless numbers, a correlation for the heat transfer from steam flow to the water jet in the steam injector was proposed.

Keywords: steam injector, next-generation reactor, steam condensation, water jet, radial heat transport, turbulent

1. Introduction

By making use of the thermal energy of steam, low-pressure gas can be pressurized and liquid can be pumped up to high elevation. The former and the latter are sometimes called as an ejector and a steam injector, respectively. These equipments do not have any moving/rotating parts. Thus, these are simple and solid in structure and reliable. A large amount of fluid can be handled even if these are small in size.

Ejectors have been used as air evacuation pumps in steam turbine systems and evaporators, and as compressors in steam jet refrigerators, and so on. Ueda [1, 2] examined the flow mechanism in the ejector and presented the design guideline of the ejectors.

Injectors have also been utilized in many areas, for example as feed water pumps in steam locomotives. Because of the advantage of the simplicity in the design and no necessity of the power to drive, Narabayashi et al. [3–5] and Iwaki et al. [6] recently examined the steam injectors by introducing the injectors into nuclear reactors as feed water pumps and safety injection pumps in mind. It has been proved that the steam injectors have the possibility that low-pressure steam can pump up water to an operating pressure of boiling water reactors (BWRs). Analytical models that can be used to design steam injectors have been also proposed.

Although steam injectors are based on proven technology and have been investigated by many researchers in the past, several things are still open to be examined. The steam injectors tested by Narabayashi et al. or Iwaki et al. were small and scaling low or scaling-up methodology should be cleared. The operating condition or range is also important. When the injectors are included in BWRs, these may experience broad conditions that may be outside design conditions occasionally. It must be clarified how the steam injector may behave under various conditions and whether there is no possibility in any condition that these may be in the way, especially in the safety aspect.

In considering the above, the most important is how to estimate the normal operating condition that the injectors function as expected and how to predict the behavior of the injectors when they go outside of the normal operating condition. These should be precisely analyzed by nuclear reactor safety analysis codes.

The essential phenomenon in the steam injectors is the conversion of the thermal energy of steam to the kinetic energy, thus the dynamic interaction and the thermal interaction between steam flow and water flow as pointed out by Iwaki et al. Fully understanding about these is required. In the present study, authors have investigated the stability of a water jet with steam condensing at the surface, the condensation heat transfer at the water jet surface, and the heat transport into the water jet for the center water jet type injectors [7–9].

In this chapter, steam condensation heat transfer to the jet surface in the steam injector was examined and the characteristics of the wavy jet surface were also reported. Additionally, the heat transfer data of the steam condensation to the water jet in the steam injector were correlated focusing on the relation between the wave motion of the jet surface and heat transport in the water jet.

2. Experimental apparatus and procedures

2.1 Experimental setup

The experimental apparatus used in the present study is schematically shown in **Figure 1**. It is composed of a steam generator, a test section, an outlet reservoir, a water tank, circulation pumps, and instruments.

The steam generator is electrically heated. It has 40 kg/h evaporative capacity at 0.5 MPa. Steam from the steam generator is superheated with ribbon heaters on piping between the steam generator and the test section and flows into the test section through an orifice flow meter. Water pumped out from the water tank also flows into the test section. The flow rate of water is measured with a rotameter. Water or water and steam mixture is collected in the outlet reservoir. Then, keeping the water level in the outlet reservoir constant, water is returned to the water tank by a pump. Steam goes back to the water tank and is discharged into the water to condense.

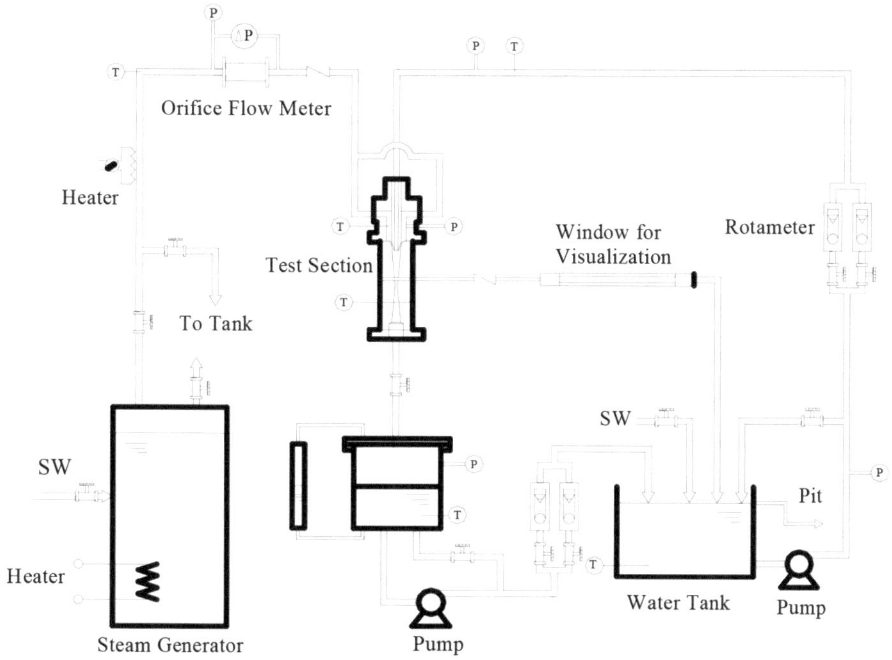

Figure 1.
Experimental apparatus. Source: Takahashi Y., Koizumi Y., Ohtake H. and Mori M., study on characteristics of thermal–hydraulic phenomena in steam injector, [internet]. Volume 4: Computational fluid dynamics, Neutronics methods and coupled codes; student paper competition. ASMEDC; 2006. Reprinted with permission.

The steam flow rate is controlled by adjusting the electric power supply to the steam generator. The water flow rate is controlled by adjusting a valve. The temperature of the water is controlled by electric heaters in the water tank and cooling coils of service water.

2.2 Test section

A water nozzle has a straight part of 125 mm in length and 5 mm in inner diameter and abruptly opens to a condensing section of steam. The water jet is blown out from this nozzle into the condensing section. Details of the test section used in the present study are illustrated in **Figure 2**.

The test section has a converging condensing section as shown in **Figure 2**. The inner diameter of the condensing section at the outlet of the water nozzle was 13.3 mm and the condensing section length was 52.9 mm. The inner diameter of the throat was 4 mm and the throat length was 5 mm. A diffuser section followed the throat. The diffuser length and the inner diameter at the outlet were 55.2 and 13.7 mm, respectively. The test section and other parts of the apparatus were well thermally insulated.

2.3 Experimental procedures

For the specified flow rate of the water jet, steam flow was supplied to the test section. During the experiment, the supplied water temperature and the water level in the outlet reservoir were kept constant. The overflow line had a check valve. When the experiment was started by supplying steam and water for the test section, the flow

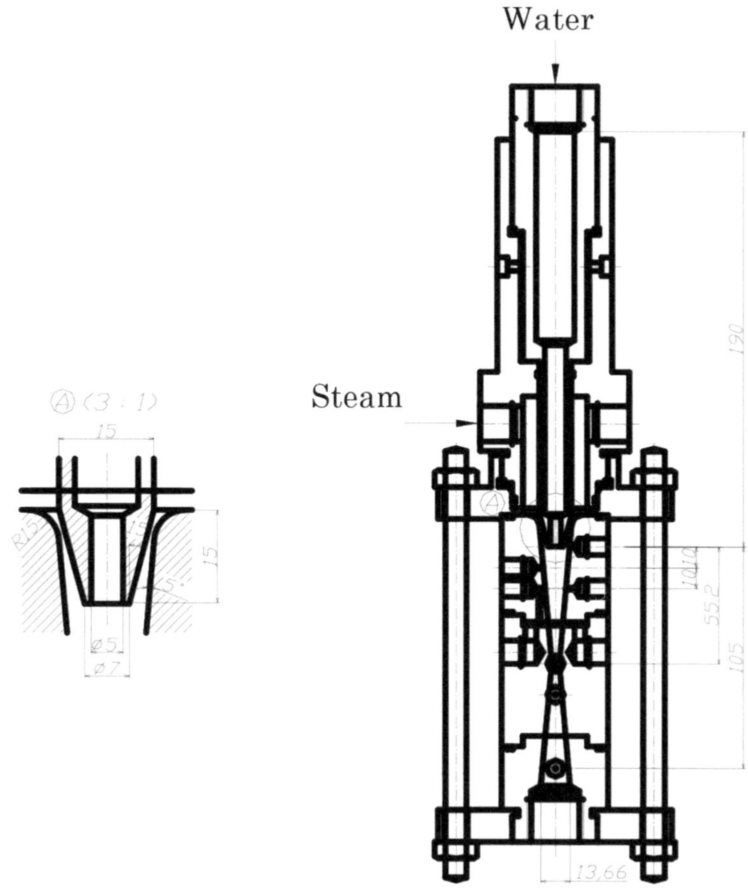

Figure 2.
Details of test section. Source: Takahashi Y., Koizumi Y., Ohtake H. and Mori M., study on characteristics of thermal–hydraulic phenomena in steam injector, [internet]. Volume 4: Computational fluid dynamics, Neutronics methods and coupled codes; student paper competition. ASMEDC; 2006. Reprinted with permission.

state in the steam injector was unstable and excess water was exhausted through the check valve in the overflow line. After the flow was stabilized, the overflow of water from the test section was stopped by the check valve. Then, the overflow line valve was manually closed. Temperature and velocity of the water jet in the test section were measured at two positions in the axial direction; at 10 and 20 mm from the outlet of the water nozzle, as shown in **Figure 2**. Pressure in the test section was also measured at similar locations. The temperature of the water jet in the mixing section was measured with an Alumel-Chromel thermocouple of 0.13 mm diameter wires. The thermocouple was radially traversed at each measuring location with an increment of 0.5 mm. The velocity of the water jet was measured with a Pitot tube of 0.8 mm diameter tube. The Pitot tube was also traversed radially in a similar way to the thermocouple.

The liquid temperature and the velocity of the water jet tested were at 20 and 35°C and from 6.8 to 17 m/s, respectively. The steam flow rate also varied from 30 to 40 kg/h in the experiments. In all conditions, the ratio of the steam to the water mass flow rate is less than 10%. In the experiments, the exit pressure of the test section was atmospheric pressure.

The flow state of the water jet in the condensing section was also visually examined. The test section for the visual experiment was made of polycarbonate and had the same as that shown in **Figure 2**. Pictures of the flow state were taken by a high-speed video camera at the flame rate of 8000 flame/s and at the shutter speed of 1/10,000 s. The pictures were recorded for one second; 8000 flames.

Parts of this chapter were originally published as a conference paper: Takahashi, Y., Koizumi Y., Ohtake H. and, Mori, M., Study on Characteristics of Thermal–Hydraulic Phenomena in Steam Injector, [Internet]. Volume 4: Computational Fluid Dynamics, Neutronics Methods and Coupled Codes; Student Paper Competition. ASMEDC; 2006. Available from: http://dx.doi.org/10.1115/ICONE14-89365.

3. Experimental and analytical results

3.1 Temperature distribution

One example of radial temperature distributions measured in the injector-type experiments is shown in **Figure 3**. The velocity of water at the nozzle outlet is 8.5 m/s

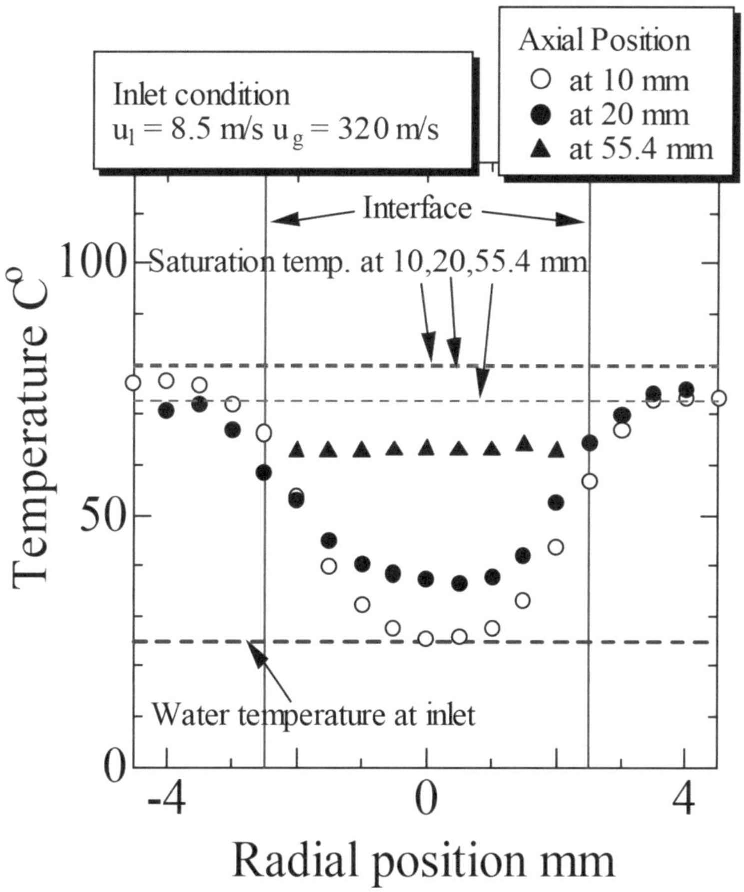

Figure 3.
Radial temperature distribution in steam injector.

and the velocity of steam at the water nozzle exit position is 320 m/s. At the position close to the nozzle outlet, 10 mm from the nozzle outlet, the temperature increase is observed only in the peripheral region of the water jet. As the flow goes downstream, at 20 mm from the nozzle outlet, the temperature increase propagates to the central region of the jet. At the position of 55.4 mm from the nozzle outlet (throat position), the radial temperature distribution becomes flat; it suggests that steam condensing has been completed until there. The temperature at the center portion of the water jet increases largely in the short distance between 10 mm and 20 mm from the nozzle outlet. The condensation of all steam flowing into the injector with the velocity of 320 m/s which corresponds to 40 kW thermal energy has completed in the very short distance of just 55 mm. It is indicated that highly efficient heat transport in the radial direction of the water jet takes place.

3.2 Velocity distribution

Measured water jet velocity distributions are illustrated in **Figure 4**. These are results for the water velocity of 8.5 m/s and the steam velocity of 320 m/s. The differential pressure measured with the Pitot tube was converted to a velocity using the density of water or the density of steam depending on the water region or the steam region, respectively. In this figure, the boundary between the water region and the steam region is expressed with a blue line. The average velocity of the water jet was derived from the measured radial velocity

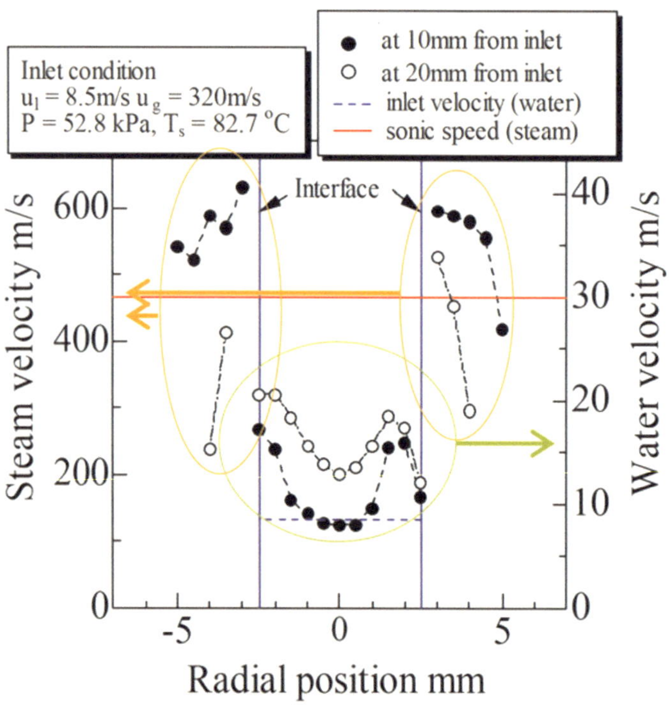

Figure 4.
Radial velocity distribution in steam injector.

distribution. Assuming that the mass flow rate of the water jet was equal to the water flow rate at the nozzle outlet, the jet surface position was obtained from the average velocity and the mass flow rate.

At 10 mm from the nozzle outlet, only the peripheral part of the jet is accelerated. At 20 mm from the nozzle outlet, the acceleration reached to the central portion of the jet, and the center part is largely accelerated. it is expected that the water jet becomes thinner as the results of the acceleration as the flows proceed downstream. However, it is not observed. It is amazing that the water jet is greatly accelerated in the very short distance of 10 mm.

In the figure, the sonic velocity of steam is illustrated. The steam velocity has reached the super-sonic velocity at 10 mm from the inlet. It suggests that steam molecules vigorously plunge into the water jet surface to condense there.

3.3 Condensation heat transfer coefficient and surface heat flux

Bulk temperature T_m is calculated from the measured liquid velocity u_l and liquid temperature T distributions as follows:

$$T_m = \frac{\int_0^{r_0} 2\pi\rho_l c_{pl} r u_l T dr}{\int_0^{r_0} 2\pi\rho_l c_{pl} r u_l dr} \tag{1}$$

The surface heat flux q_s of steam condensation to the jet surface can be related to the increasing rate of bulk temperature to the flow direction x as:

$$\frac{dT_m}{dx} = \frac{\pi D q_s}{c_{pl} m_l} \tag{2}$$

where D is the water jet diameter and m_l is the water jet flow rate. The condensation heat transfer coefficient is defined by using the local water subcooling that is defined by using the steam saturation temperature for pressure at the measuring position and the bulk water temperature at the measured position as follows;

$$h = \frac{q_s}{T_{sat} - T_m} \tag{3}$$

The heat fluxes q_s of steam condensation to the jet surface derived with Eqs. (1) and (2) are presented in **Figure 5**. The horizontal axis is the local water subcooling. It was expected from **Figure 5** that the condensation heat transfer coefficient would show decreasing trend for the water subcooling since the surface heat flux seems to be constant with an increase in the subcooling and the heat transfer coefficient was in inverse proportion to the subcooling; Eq. (3).

Heat transfer coefficients h derived by Eq. (3) in the experiments are plotted for the local water subcooling in **Figure 6**. Measured heat transfer coefficients express a weakly decreasing trend for an increase in the inlet liquid subcooling. Those are much lower than the ideal condensation heat transfer coefficient [10].

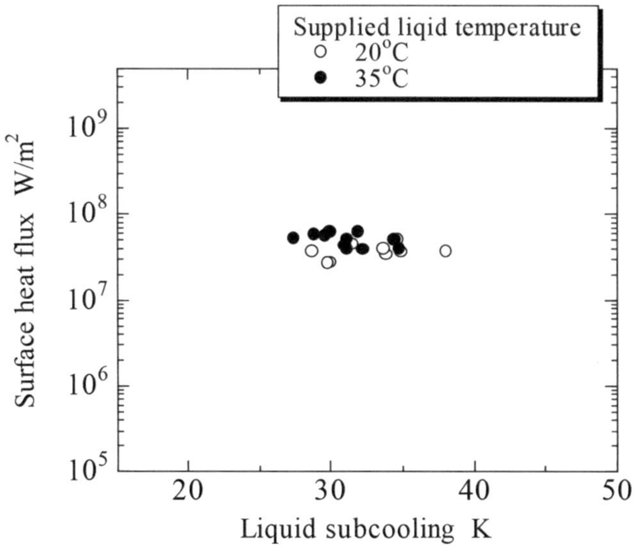

Figure 5.
Surface heat flux.

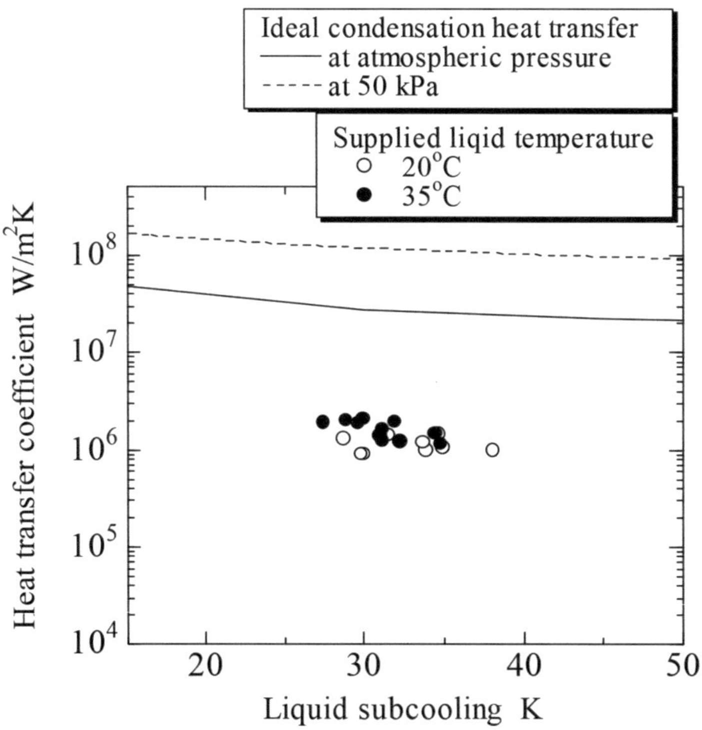

Figure 6.
Heat transfer coefficient.

Figure 7 shows the Re-Nu correlation that was obtained from this experiment. The value of Nu is one or two orders as large as the Dittus-Boelter correlation. This result clearly expresses that tremendously effective heat transfer was done in the condensation area.

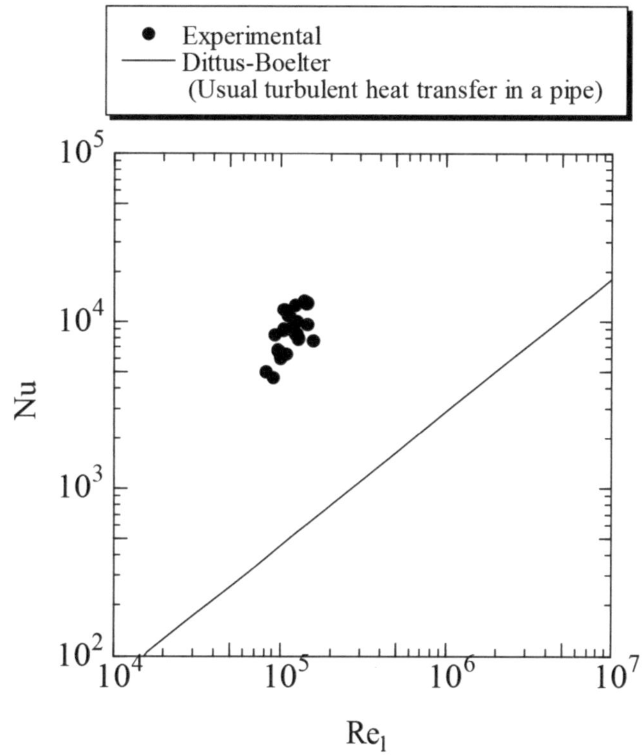

Figure 7.
Re-nu correlation.

3.4 Visualization

Figure 8 shows the image of the water jet in the steam injector taken by high-speed video camera. In **Figure 8**(a), only water flows, and steam is not supplied. The original image on the left is binarized on the right. Although there are tiny waves on the surface, the shape of the water jet is kept round and straight. When steam was provided and the steam injector functioned as a pump, the jet surface looks like the water jet is foamy; tiny vapor bubbles are dissolved into the water jet (**Figure 8**(b)). And it is noted that there is large-clear wave motion on the water jet surface.

From these results, it was supposed that the wavy motion on the water jet surface created the effective large-internal circulation flow in the water jet and the tremendously effective heat transport into the center portion of the water jet.

From the pictures of the water jet surface, the characteristics of waves on the surface; the wave height, the wave length, and the wave velocity, were obtained. A total of 50 large waves were randomly selected in the recorded pictures and then the wave heights were measured. The average value of these was defined as the wave height. Similarly, 50 large waves were randomly selected, and then traveling distance in a certain time period was obtained. From these, the wave velocities were calculated. The average of these was also defined as the wave velocities in the present experiments.

Figure 9 shows one example of the characteristics of the wave on the jet surface obtained in the present study. The water temperature was 20 and 35°C. The wavelength was almost constant regardless of the velocity difference between the water jet

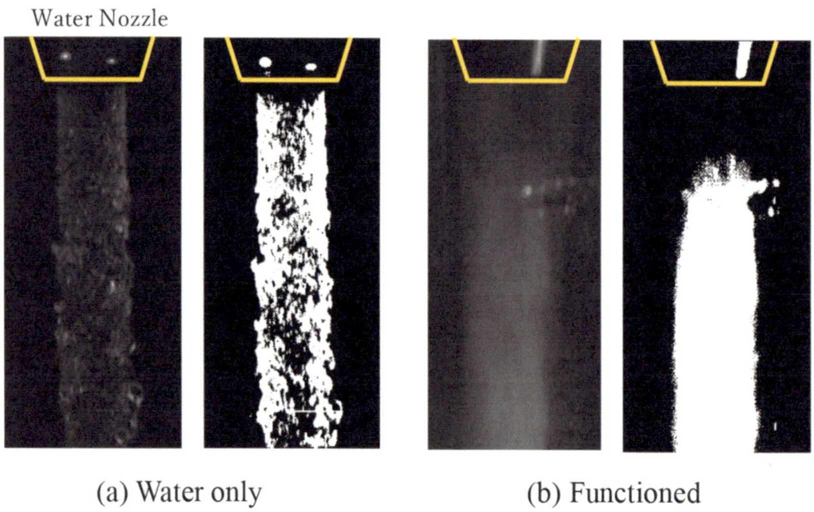

Figure 8.
Water jet behavior.

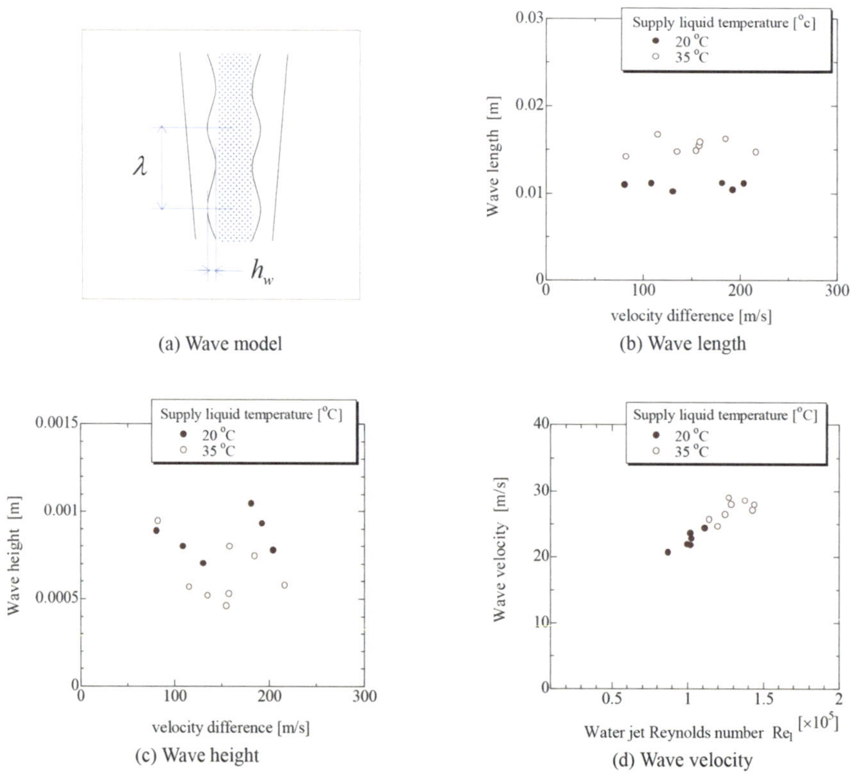

Figure 9.
Wave characteristics.

and the steam flow. The wave velocities indicate the tendency to increase with the water jet Reynolds number. The wave velocity is in the range from 20 to 30 m/s. The average water jet velocity was in the range from 15 to 20 m/s.

4. Correlation of dimensionless numbers

4.1 Non-dimensional analysis

The dimensionless numbers were derived from the non-dimensional analysis. The obtained dimensionless numbers are;

The Nusselt number: $\mathbf{Nu} = hD/k_{l.}$

The water jet Reynolds number: $\mathbf{Re}_{\mathbf{l}} = u_l D/\nu_l$

The Prandtl number: $\mathbf{Pr} = \rho_l \nu_l c_{pl}/k_l$

The Weber number is defined by the difference between the average steam velocity and the average jet velocity Δu and the jet diameter: $\mathbf{We}_{\mathbf{\Delta u_l}} = \rho_l \Delta u^2 D/\sigma$

The Froude number: $\mathbf{Fr} = {u_w}^2/gD$

The Reynolds number defined by the velocity difference between the jet and the steam flow Δu: $\mathbf{Re}_{\mathbf{\Delta ul}} = \Delta u D/\nu_l$

The Reynolds number is defined by the wavelength λ and the differential velocity between the jet and the steam flow Δu: $\mathbf{Re}_{\mathbf{\Delta u\lambda}} = \Delta u \lambda/\nu$

The Reynolds number defined by the wave height h_W and the differential velocity between the jet and the steam flow Δu: $\mathbf{Re}_{\mathbf{\Delta uh}} = \Delta u h_W/\nu$

Non-dimensional wave velocity: $\mathbf{N}_{\mathbf{uw}} = \nu g/{u_W}^3$

Combinations of Dimensionless Number: $\mathbf{We}_{\mathbf{\Delta u}}{}^{-1} \cdot \mathbf{Re}_{\mathbf{\Delta ul}}$, $\mathbf{Nu} \cdot \mathbf{Re}_{\mathbf{l}}{}^{-1}$

4.2 Heat transfer correlation

By using some of the non-dimensional numbers that were derived by the non-dimensional analysis, the best-fit correlation for the steam condensation heat transfer to the water jet in the steam injector was developed.

In **Figure 10**, the relation between the non-dimensional parameter groups is presented. Some trend is noticed. In the present experiments, the temperature of the

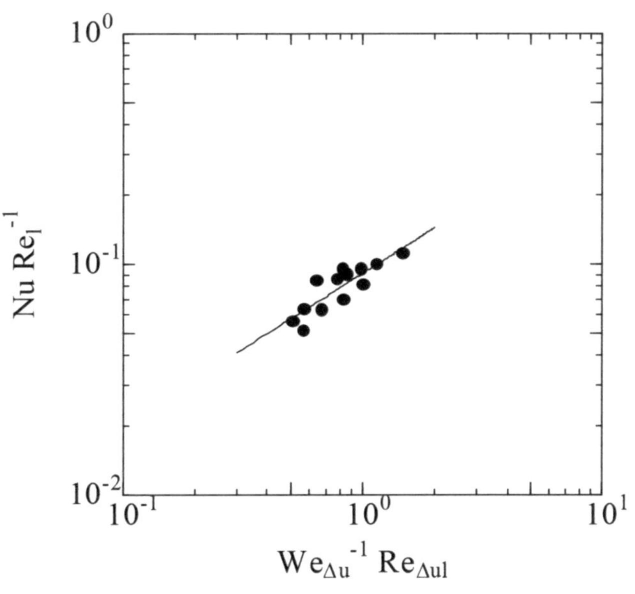

Figure 10.
Dimensionless number combination.

water jet was varied. In order to check the effect of the physical properties, the dependency of the heat transfer on the Prandtl number is presented in **Figure 11**. The clear dependency of the heat transfer on the Prandtl number is noticed.

Finally, the best-fit correlation for the steam condensation heat transfer in the steam injector is developed as

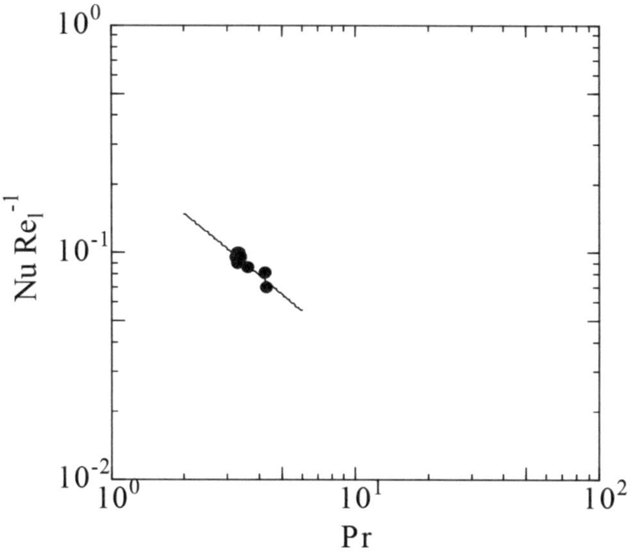

Figure 11.
Influence of Prandtl number.

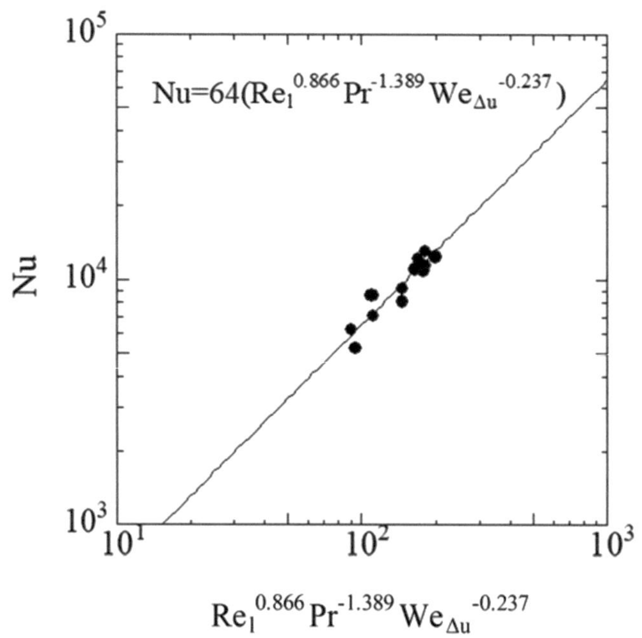

Figure 12.
Heat transfer correlation.

$$\mathbf{Nu} = 64\left(\boldsymbol{Re}^{\,0.866}\mathbf{Pr}^{-1.389}\mathbf{We}_{\boldsymbol{\Delta}\mathbf{u}}{}^{-0.237}\right) \tag{4}$$

As shown in **Figure 12**, the agreement between the experimental results and the proposed correlation; Eq. (4) is quite well.

5. Conclusions

Steam condensation heat transfer to a water jet in a steam injector was examined. Following conclusions were obtained.

1. The measured velocity distribution exhibited that the velocity of steam around the water jet was super-sonic velocity and the water velocity at the peripheral region was considerably faster than that in the central region. It suggested that the water jet was greatly accelerated by the steam flow around the jet. The radial and the axial temperature distributions expressed that the water temperature at the central region jumped up in a short distance. It implied that considerably effective-radial heat transport took place in the water jet.

2. The heat transfer coefficient showed a slightly decreasing trend for water subcooling and the condensation heat flux was almost independent of the subcooling. The heat transfer coefficient of steam condensation to the water jet surface was much higher than that of the turbulent heat transfer in the circular tube.

3. As a result of observation, it was clarified that the interface between the water jet and the steam flow was very wavy. It was supposed that the wavy motion on the water jet surface created tremendously effective heat transport into the center portion of the water jet. The wavelength, wave height, and wave velocity were measured from pictures taken by a high-speed video camera.

4. From the non-dimension analysis and the comparison with the experimental results, the heat transfer correlation of the jet flow accompanying the direct condensation of steam on the surface in the steam injector was proposed.

Acknowledgements

The present research project has been carried out by Tokyo Electric Power Company, Toshiba Corporation, and six Universities in Japan, funded by the Institute of Applied Energy (IAE) of Japan as the national public research-funded program.

Nomenclature

c_p	Specific heat [J/kg·K]
D	Water jet diameter [m]
g	Gravitational acceleration [m/s^2]
h	Heat transfer coefficient [W/m^2·K]

h_W	Wave height [m]
k	Thermal conductivity [W/m·K]
m	Mass flow rate [kg/s]
P	Pressure [Pa]
q	Heat flux [W/m^2]
r	Radial position [m]
T	Temperature [K], [°C]
u	Velocity [m/s]
x	Axial position [m]
y	Radial position from wall [m]

Greek Symbols

Δu	Difference between average steam velocity and average jet velocity [m/s]
λ	Wave length [m]
ν	Kinematic viscosity [m^2/s]
ρ	Density [kg/m^3]
σ	Surface tension [N/m]

Subscripts

g	Gas
l	Liquid
m	Mean
s	Surface
sat	Saturation
W	Wave

Author details

Yasuo Koizumi
The University of Electro-Communications, Chofu, Japan

*Address all correspondence to: koizumiy@shinshu-u.ac.jp

References

[1] Ueda T. Study on the steam ejection (1st report). Transactions of the JSME. 1952;**18**(67):97-103

[2] Ueda T. Study on the steam ejection (2nd report). Transactions of the JSME. 1952;**18**(67):103-111

[3] Narabayashi T, Nei H, Ozaki O, Shioiri A, Mizumachi W. Study on high-performance steam injector (1st report, development of analytical model for characteristic evaluation). Transactions of the JSME. 1996;**62B**(597):1833-1840

[4] Narabayashi T, Mizumachi W, Mori M. Study on two-phase flow dynamics in steam injectors. Nuclear Engineering Design. 1997;**175**:147-156

[5] Narabayashi T, Mori M, Nakamura M, Ohmori S. Study on two-phase flow dynamics in steam injectors II. Nuclear Engineering Design. 2000; **200**:261-271

[6] Iwaki C, Narabayashi T, Mori M, Ohmori S. Study on high-performance steam injector (2nd report, measurement of jet structure). Transactions of the JSME. 2003;**69B**(684):1814-1821

[7] Koizumi Y, Ohtake H, Yamashita N, Miyashita T, Mori M. Development of technologies on innovative simplified nuclear power plant using high efficiency steam injectors (6) direct condensation heat transfer and stability of water jet in steam injector. In: ASME-JSME-CNS 13th International Conference on Nuclear Engineering; Beijing, China. New York, US: ASME; 2005

[8] Takahashi Y, Koizumi Y, Ohtake H, Yamashita N, Miyashita T, Mori M. Study on direct condensation heat transfer in steam injector. Progress in Multiphase Flow Research. 2006;**I**:241-248

[9] Takahashi Y, Koizumi Y, Ohtake, Mori M. Study on characteristics of thermal-hydraulic phenomena in steam injector. In: ASME-JSME-CNS 14th International Conference on Nuclear Engineering; Miami, Florida, USA. New York, US: ASME; 2006

[10] Isshiki N. Heat Transfer Engineering. Tokyo, Japan: Morikitashuppan Co.; 1967. pp. 133-135